Praise for
Beyond the Root Cellar

"As a seasoned practitioner and teacher of market gardening, it's rare for a book to ignite my passion, but Sam's work does just that. Overflowing with invaluable tips and innovative strategies, this guide goes beyond the typical fast crops seen in most market gardens. It dives deep into storage vegetables for year-round growing, empowering growers to extend their season. In the face of constraints, we discover our greatest teachers, and Sam's insights from Alaska and its short growing season have broadened his perspective. This new book has certainly broadened mine, and I encourage all growers to study it and level up their farming game."

— **Jean-Martin Fortier**, author of *The Market Gardener* and *The Winter Market Gardener*

"*Beyond the Root Cellar* is the best resource on storing vegetables that I know of. This is next-level market gardening. Knapp is practical and inspiring—he shows modern methods for storing food and makes a strong case for why you should probably be storing more food on your own farm. Storing food is a win for customers, for farmers, and for the planet. If you grow vegetables, I recommend that you pick up a copy of this book."

— **Ben Hartman**, author of *The Lean Micro Farm*

"This book is the next step in creating resilient food systems and communities, telling us how to provide high-quality, nutritious local vegetables all year rather than importing them from warmer places in the winter. It is a vital resource for small farmers and home gardeners who want to extend their growing and selling season. Sam presents extensive and detailed information about the best varieties for long-term winter storage; which crops to harvest first when a fall cold snap is predicted and everything needs to be done at once; how to prepare crops for storage with curing, precooling, and proper washing; appropriate storage containers; and optimal storage conditions for each crop. I wish I had read this before I designed my own root cellar!"

— **Helen Atthowe**, author of *The Ecological Farm*

"*Beyond the Root Cellar* is a must-read for anyone who wants to store produce for any length of time. It's too much work and investment to grow produce, only for it to decline in quality in storage. Produce can rot surprisingly quickly when stored under the wrong conditions, but it can also last for a surprisingly long time under the right conditions. Sam Knapp shows us the reality that vegetables will last only if harvested at the right stage, with the proper postharvest handling, temperature, humidity, and other factors. Detailing everything from the produce itself to the infrastructure you need to store it and the equipment that will help you maintain the right conditions, this book is an important contribution to making the entire year local food season! Even if you don't plan on storing produce all winter, this book will show you how to keep it as fresh as possible for as long as possible."

— **Andrew Mefferd**, editor, *Growing for Market* magazine, author of *The Greenhouse and Hoophouse Grower's Handbook*

"*Beyond the Root Cellar* is a very thorough exposition of how to build, stock, and manage a root cellar. While it is definitely most helpful to large commercial operations that are mechanized, it goes into great detail across the board. Sam Kapp lays out the proper crops to store in a root cellar, along with their varying temperature and humidity needs. Because he lives in Alaska, it is particularly impressive what he has done to raise high-quality produce and then store it for later sale. An impressive and highly detailed book!"

— **Julie Rawson**, organic farmer; author of *Many Hands Make a Farm*

Beyond the
Root Cellar

Beyond the Root Cellar

The Market Gardener's Guide
to Growing and Storing
Vegetables for Off-Season
Sales and Food Security

Sam Knapp

Chelsea Green Publishing
White River Junction, Vermont
London, UK

Copyright © 2024 by Sam Knapp.
All rights reserved.

Unless otherwise noted, all photographs copyright © 2024 by Sam Knapp.
Photograph on title page courtesy of the Alaska Farmers Market Association and Kyra Harty.

No part of this book may be transmitted or reproduced in any form by any means without permission in writing from the publisher.

Developmental Editor: Natalie Wallace
Copy Editor: Will Solomon
Proofreader: Fran Pulver
Indexer: Shana Milkie
Designer: Melissa Jacobson

Printed in the United States of America.
First printing November 2024.
10 9 8 7 6 5 4 3 2 1   24 25 26 27 28

Our Commitment to Green Publishing

Chelsea Green sees publishing as a tool for cultural change and ecological stewardship. We strive to align our book manufacturing practices with our editorial mission and to reduce the impact of our business enterprise in the environment. We print our books using vegetable-based inks whenever possible. This book may cost slightly more because it was printed on paper from responsibly managed forests, and we hope you'll agree that it's worth it. *Beyond the Root Cellar* was printed on paper supplied by Versa that is certified by the Forest Stewardship Council.®

Library of Congress Cataloging-in-Publication Data

Names: Knapp, Sam, 1990– author.
Title: Beyond the root cellar : the market gardener's guide to growing and storing vegetables for off-season sales and food security / Sam Knapp.
Description: White River Junction, Vermont : Chelsea Green Publishing, 2024. | Includes bibliographical references and index.
Identifiers: LCCN 2024038309 (print) | LCCN 2024038310 (ebook) | ISBN 9781645022107 (paperback) | ISBN 9781645022114 (ebook)
Subjects: LCSH: Vegetables—Storage—Handbooks, manuals, etc. | Truck farming—Handbooks, manuals, etc. | Cold storage—Handbooks, manuals, etc. | Farm produce—Storage—Handbooks, manuals, etc. | Handbooks and manuals.
Classification: LCC SB351.R65 K53 2024 (print) | LCC SB351.R65 (ebook) | DDC 641.4—dc23/eng/20240925
LC record available at https://lccn.loc.gov/2024038309
LC ebook record available at https://lccn.loc.gov/2024038310

Chelsea Green Publishing
White River Junction, Vermont, USA
London, UK
www.chelseagreen.com

Contents

Introduction

Sometimes, life presents you with conundrums. When I was twenty-six years old, it presented me with one. Here's some context. I had been out of college for several years and had caught the farming bug. Over a handful of years, I had worked for other farmers, walking a meandering path that led to a stint managing a large community garden and farmers market as an AmeriCorps volunteer. To join AmeriCorps, I had moved several hours away from Danielle, my girlfriend and now wife, and I hoped to rejoin her in the Upper Peninsula of Michigan, where she was going to graduate school. Near the end of my one-year service term, I got an opportunity to return to Michigan as a paid graduate student in plant ecology. While this was great news, I was reluctant to take a hiatus from farming, so I decided to find work on a nearby vegetable farm alongside school. Problem was, there were no nearby vegetable farms hiring workers. In fact, there were only a handful of vegetable farms in that entire region of the Upper Peninsula. My conundrum led me toward two life-altering decisions.

The first important decision was that I would run a farm for myself. Nothing huge. Something large enough that I would feel like a real farmer but small enough that I could manage the workload on top of school. The small house Danielle and I rented was part of a 120-acre cattle farm, and the owners were happy to let me use a small, quarter-acre plot to grow vegetables. It would be an incubator farm of sorts; the land rental was, for all practical purposes, free, and I could put all the income from the farm into equipment and savings. As I developed a plan for the farm, however, I realized there were many obstacles—of those, time was the biggest.

During that first growing season, I would still be living and working several hours away. While I could drive up every week or two to tend the farm over the weekends, I couldn't be there consistently to shepherd needy crops or to prep for and attend farmers markets. That problem would continue past the first year, as I had obligations as a graduate student to do field research all over the Upper Peninsula and northern Wisconsin during the summers. These commitments eliminated any possibility of running a traditional market or CSA farm. I simply didn't have time. I began looking into winter-oriented farming, as I was interested in the concept of year-round greenhouse production. However, as a cash-strapped graduate student on temporarily rented land, I didn't have the resources or the desire to invest heavily in greenhouse infrastructure. I needed to find another way.

At the time, I was participating in the Farmer-to-Farmer Mentorship Program organized by the Midwest Organic and Sustainable Education Service (now Marbleseed). My farmer mentor, Brian Gronske of Groché Organic Farms, suggested what became the second life-altering decision: organizing a winter CSA with storage crops. I had never heard of such a thing, but the more we talked, the more it made sense. Storage crops would take minimal time to plant and tend during the summer, and all the work of packing and selling could happen during the winter when I had more time. Furthermore, it would be relatively easy to market my vegetables during the winter when other regional farms weren't selling. Brian helped me craft the initial crop plans and offered invaluable financial advice. Unfortunately, he didn't store crops commercially on his farm, so I needed to turn elsewhere to learn the fine details of long-term winter storage. I plowed ahead and started what became Root Cellar Farm.

Despite my trepidation and the steep learning curve, that first season went well. The small cold-storage space I'd built in an auxiliary basement stairwell worked admirably, and because I grew a diversity of crops, the operation was resilient against several of my rookie mistakes. That said, the first season at Root Cellar Farm was terrifying. I rarely felt confident that my methods of preparing and storing crops would work. I had pieced together information from books, university extensions, and other farmers, but most of these resources were aimed at home gardeners and homesteaders using traditional root cellars. I have nothing against these time-tested methods; they're the same ones many of our ancestors used to feed themselves through the winter. Some advice seemed impractical for commercial scales, though. Was I really supposed to store thousands of pounds of root vegetables in damp sand? There was also conflicting information. For example, some sources were adamant to avoid washing veggies before storage while others advocated for it. Could they both be right?

Apart from my storage methods actually working, what surprised me most was the excitement the farm generated. Because I planned conservatively, I had excess of some crops beyond my needs for the CSA. I sold these extras to a nearby co-op grocery that carried local farm products. The manager told me that, because other area farms ran out of produce by December, anything I could provide beyond then would be a boon for them. Word about my small, ten-member CSA also spread quickly, and I soon had a long wait list. I realized I was tapping into something big. The customer base that bought local products and made the summer farmers market thrive didn't go away come wintertime. Faced with the dearth of local veggies during offseason months, they were hungrier than ever for great-tasting, locally grown produce. Whatever I could grow and store deep into winter would not only sell but sell with gusto.

I operated Root Cellar Farm from our rented land and house in Toivola, Michigan, for three seasons, and loved every moment of it. By the third year, I had adjusted my crop mix to improve yields, had expanded to cultivate about one third of an acre, and had increased membership in the winter CSA to twenty families. Even so, I barely scratched the surface of the wintertime demand. During that time, Danielle and I got married just down the road from the farm. We fed the guests at our reception with veggies mostly from Root Cellar Farm!

I would have loved to stay and continue expanding the farm, but alas, it wasn't meant to be. Danielle and I had both graduated, and it was time to move on. Years earlier, we had fallen in love with life in the far north, and we heeded the siren call to return. Danielle focused her job search in Alaska, eventually accepting a job with the National Park Service in Fairbanks. Luckily, I knew something of the farming scene in Fairbanks, having worked there for a summer years earlier on a small vegetable farm. My rough plan was to find a job while starting a storage-focused farm on the side.

Moving to and getting settled in Alaska was a saga that I won't recount here. Suffice it to say that driving there in January was both terrifying and exhilarating; we lost more and more daylight as we traveled north, and our car thermometer read −40°F more than once. Driving the Alaska-Canadian Highway again in May—with a trailer full of farm equipment and household belongings and a pickup bed filled with chickens—was less daunting but just as long. I'll never forget the looks I got while pumping gas as roosters crowed inside the truck topper.

We bought a house on three sloped and forested acres soon after arriving. Obviously, this wasn't the best choice for starting a farm, but cleared land comes at a premium around Fairbanks, and more importantly, the property had a well (which isn't common). After the success of Root Cellar Farm in the Upper Peninsula, I suspected a similar farm would flourish in Fairbanks, so I started investigating. I asked around among my farmer contacts and perused the produce

Starting a new farm in a new place is a challenge, with new things to learn but new friends as well.

sections of local groceries. Even in January, there weren't any local veggies other than potatoes. The summer farmers market was wildly popular, but even the largest diversified veggie farms sold out their inventories by early December. It was clear that the winter marketplace around Fairbanks was primed for a storage-oriented farm. All I had to do was build it.

Anyone who has built a farm knows that it's not an easy task and requires deep commitments of your time, emotions, and finances. Importantly, building a farm also takes knowledge. While I had several years of experience from Root Cellar Farm, I hadn't built it from scratch. I had much to learn and many new hats to wear. In my research, I was often unable to find resources tailored to my needs. It was frustrating, because I knew the information was out there somewhere; plenty of other farmers had built successful winter storage facilities, yet there I was reinventing the wheel. Again, I plowed ahead and hoped for the best.

Eventually, we were able to get Offbeet Farm off the ground. We secured financing and cleared a portion of

our land before constructing a building tailored for winter vegetable storage and processing. Our location in interior Alaska brought some special challenges, but it brought opportunities as well. As in Michigan, the farm quickly generated a lot of excitement, and even though the farm is larger, our production is still dwarfed by the wintertime demand. The reason I originally gravitated toward storage farming—to spread the workload more evenly throughout the year—still applies at the larger scale. My summers don't have that frantic pace so common on traditional veggie farms because the time-consuming work of prepping and selling produce happens during fall and winter. The farm is still a work in progress, but we've reached a level of stability and success to refine our systems and look toward the future.

As I've already hinted, winter storage has many potential benefits for farmers, whether it's the focus of your operation or a side project. Winter storage allows farmers at scales large and small to generate additional income, shunt some of the workload to the offseason, and provide their communities with local food during winter months. Dabbling into winter storage might help some farms retain customers by keeping them engaged throughout the year. Farms might also retain good employees with offseason packing work. There may even be larger societal benefits from winter storage, like bolstering local food systems and raising public confidence in the ability of local farms to provide.

If storage farming has so much potential, why aren't more people doing it? I think there's a negative feedback loop that keeps many farmers from branching out into storage. There are high upfront costs for infrastructure, handling equipment, and storage containers. Storage crops aren't those high-dollar darlings that farmers gravitate toward to generate income. Storage farming also carries more risk; there's risk involved not only in producing a crop, but also in storing it. Lastly, and perhaps most importantly, relatively few farmers currently store crops into winter. Because most farmers learn the trade by working for others, there aren't many new farmers entering the scene with hands-on storage experience.

There are several reasons why I decided to write this book, but an important one is to break this cycle. Farmers need a resource for learning the craft of winter vegetable storage, and they need examples of other farmers already succeeding at it. Looking back at my own journey into winter storage, I sorely wish this book had existed—not only to alleviate my anxiety and save me from mistakes, but to show me what's possible. This book is a collection of the knowledge I wish I'd had when I started Root Cellar Farm and later built Offbeet Farm. I hope it inspires more farmers, both new and experienced, to try storing crops commercially. I also hope it gives gardeners and homesteaders a window into modern methods of food storage. Long-term storage in climate-controlled facilities is not just for industrial-scale farms, and this book is meant to show growers how to succeed with storage at scales large and small. Alongside collected knowledge from universities and research institutions,

I've compiled the experiences of farmers. My own farming mentors saved me from years of mistakes, and I hope you can benefit from the decades of their collected observations, trial and error, and knowledge.

The book is broken into four parts to guide your reading. Part one is a hands-on guide to getting crops from the field and into storage in the best shape possible. From variety selection to harvesting at peak maturity, from curing and washing to placing veggies into storage containers, these chapters cover practical details any grower should consider when readying crops for winter storage. I mostly leave planting and tending up to you; this book is, instead, meant to help with the steps that come after.

Part two is about designing the structures and practices that make a storage farm successful. These chapters cover some aspects of storage farming as a business, as well as building functional and efficient storage spaces with help from modern building materials and technology. This section is not a construction guide. Instead, some of the chapters will help you have informed conversations with contractors or, if you have the skills, choose the right materials and designs to build great storage spaces yourself.

Part three is a detailed listing of eighteen crops amenable to winter storage. I've delved through stacks of scientific literature and compiled farmers' knowledge to provide tested methods of harvesting, processing, and storing that work for these individual crops. If you need specific information on a specific crop, start here. The final part of the book is a collection of profiles from farms already storing vegetables commercially. I interviewed storage farmers across wintery North America (and one in North Carolina!) to hear their stories and to learn how they each harvest, process, store, and sell. Take this opportunity to learn from their mistakes, their successes, and their advice.

I sincerely hope you find this book useful and feel inspired about starting your own journey into wintertime storage. The more of us there are growing and storing locally sourced food into winter, the better off our communities and food systems will be.

Part 1

From Field to Storage

Carrots are one of the most popular and recognizable of today's biennial crops. *Photo courtesy of Phil Knapp.*

Chapter 1

Choosing Crops and Varieties

Before you ever pull vegetables out of storage, harvest from the field, or even plant the seeds, you have to choose which crops and varieties you're growing. These choices are key for any farmer, but even more so for farmers storing vegetables into the winter. No matter how well you time your harvests or how carefully you prepare for storage, poor crop or variety choices can derail all your efforts. That's because not all crops are suited for storage, and for those that are, some cultivars are better than others.

From the outset, you should ask yourself two questions: "Can I grow this?" and "Will this store for the duration I need it to?" The first question should apply to everything you grow, not just produce destined for storage. You should be reasonably sure you can grow a particular crop or variety to healthy maturity given your local climate, soils, infrastructure, and pest/disease pressures. The second question applies specifically to what you plan to store before selling. As with growing a crop, you want reasonable certainty you can store it successfully, too. Whatever the cultivar, it should maintain premium quality for the duration you need it to in the conditions you can provide. How do you know this with certainty? Estimating storage life is part of the learning curve of storage farming. While there are general rules of thumb—see part 3 for more details—each variety is different. You can gain more insights from the descriptions in seed catalogs, advice from other farmers, and—best of all—experience on your own farm.

You are probably aware of the common storage crops—the roots, tubers, leaves, stems, and fruits commonly sold at groceries and markets each fall and winter. Even though some behave similarly in storage, today's common storage crops come from many botanical families and have a variety of life cycles. All of these factors affect their behavior and longevity in storage. While not necessary for success, knowledge of their origins and physiologies can help to guide your decisions as you peruse seed catalogs each year.

Storage Crops and Life Cycles

Many of the current-day storage crops are biennials, meaning they produce flowers and seeds in their second, rather than first, growing season. However, most of today's domesticated biennial crops were developed from ancestor plants with annual life cycles. The first domesticated biennials appeared around 6,000 years ago.[1] As people selectively bred for larger edible plant parts—be those roots, stems, or leaves—the plants put more time and energy into vegetative growth and eventually lost their ability to reproduce in a single growing season. This forced them to survive through winter to produce flowers and seeds the following year. This capacity for survival turned the plants into reliable food sources during the winter months, which incentivized people to keep growing and developing them as crops.

Common biennial crops today include those from the parsley family (Apiaceae) like carrots, parsnips, and celeriac; beets from the goosefoot family (Chenopodiaceae); onions, garlic, leeks, and shallots from the lily family (Liliaceae); and cabbages, kohlrabis, winter radishes, rutabagas, turnips, and Brussels sprouts from the mustard family (Brassicaceae). Each of these plants has a storage organ that stores energy and protects the meristem, which sends forth flowering parts during the second growing season. These storage organs are also the parts we eat.

Even though we call many of these crops root vegetables, most of the storage organs aren't technically roots. In fact, carrots and parsnips are the only two taproots we commonly eat (less-common taproot crops include parsley root—a.k.a. arat—salsify, and burdock root). The storage organs of beets, celeriac, radishes, rutabagas, and turnips are fused portions of root and stem known as hypocotyls. The bulbs of common alliums like onions, garlic, and shallots are not roots but rather swollen leaf bases growing from dwarf stems. Likewise, the culinary portions of leeks are tightly bundled leaves known as pseudostems. Cabbages and Brussels sprouts have storage organs consisting of tightly overlapping leaves protecting the cores, where the apical meristems for future flowering shoots reside.

Storage organs use several strategies to survive winter and protect the meristems. One strategy is

The tightly packed and overlapping leaves of a cabbage protect the apical meristem at the tip of the core from damage and cold temperatures.

upping their cold hardiness through a process called cold acclimation. When exposed to cooling temperatures and decreasing daylength, some plants make themselves more tolerant to cold by, among other things, producing soluble sugars and alcohols, which lower their freezing points, and making special proteins called ice-binding proteins, which reduce the size of ice crystals forming inside the plants.[2] These proteins are the same type as those found in certain fish, insects, and amphibians that survive freezing solid during the winter, and they allow some tundra plants to survive temperatures dipping below −60°F (−51°C). Another survival strategy is dormancy, which is a lowered metabolic state activated either by plant hormones or environmental conditions such as low temperatures. Cabbages and onions are examples of biennial crops that enter true, hormonally controlled dormancy, while many other biennial crops, including carrots, beets, and parsnips, are dormant only when exposed to lowered temperatures and decreased daylength.[3]

Not all storage crops have biennial life cycles. Winter squashes and pumpkins from the gourd family (Cucurbitaceae) are true annuals and produce flowers and seeds during their first and only growing season. Their fortitude in storage comes from slow senescence (decay) of fully matured fruits. This slow decay may have been an evolutionary strategy favoring prolonged periods for animals to eat the fruits and spread their seeds. People selectively bred squashes and pumpkins for storage by keeping and planting those that tasted best and stored longest—usually the ones with protective, waxy rinds and lots of orange carotenoids in their flesh.

Several other common storage crops are perennials grown as annuals. These include potatoes from the nightshade family (Solanaceae), sweet potatoes from the morning glory family (Convolvulaceae), and sunchokes (a.k.a. Jerusalem artichokes) from the daisy family (Asteraceae). I'm intentionally leaving out true yams (from Dioscoreaceae, the yam family) because they're not commonly grown outside the tropics. Depending on your local climate, potatoes, sweet potatoes, and sunchokes might readily survive as perennials if left in the ground to overwinter. If you've ever seen volunteer potato plants in your garden, you know this is true. As annuals, these crops also produce flowers and seeds in a single growing season.

The storage organs for these three perennial crops are tubers: stem tubers for potatoes and sunchokes and root tubers for sweet potatoes. Tubers are sections of roots or stems specialized for storing nutrients. The parent species of these crops already had tubers adapted to survive winters and droughts by going dormant until growing conditions improved. People selectively bred these perennials to produce larger, starchier tubers and, for potatoes, to remove toxins present in the parent species.

Dormancy in Storage

Crops that do well in long-term winter storage almost all enter states of dormancy, meaning they reduce their metabolic activity to base levels and suppress sprouting in their shoots and roots. Our goal as storage farmers is to provide storage conditions that keep crops dormant while also preventing chilling damage and decay. Most crops enter dormancy due to some combination of decreasing temperatures, decreasing daylength, or dieback of the parent plant. The specific conditions that maintain dormancy probably relate to the climate where each crop originated. For example, both winter squashes and sweet potatoes remain dormant between about 50 and 60°F (10 and 16°C), and both originated in the tropics of Central America and northern South America.[4] Carrots, on the other hand, were domesticated in both Afghanistan and Turkey before spreading to the rest of Asia and Europe, and they require cooler temps to remain dormant, preferably around 32°F (0°C).[5]

With regard to crops in storage, dormancy comes in two flavors: environmentally controlled (induced) and hormonally controlled. Generally speaking, crops with environmentally controlled dormancy will resprout if you expose them to warm temperatures and moisture.

A turnip sprouts new leaves in storage as it begins to break its dormancy.

Crops with hormonally controlled dormancy will not resprout when dormant, regardless of the surrounding conditions. Triggers for the end of hormonal dormancy vary by crop and even cultivar. It's often a combination of temperature and time (in other words, cumulative exposure) that triggers the production of hormones to kickstart metabolism and regrowth.

As crops break dormancy, carbohydrates begin changing form and moving to where they're needed to fuel growth. Sprouting of both roots and leaves and loss of sweetness are all signs that a crop is breaking dormancy, both hormonal and induced. When this happens, it's bad news for commercial storage farmers because sprouting crops quickly become unsalable.

Potatoes and sweet potatoes are examples of similar crops with different types of dormancies. Dormancy in sweet potato tubers is environmentally controlled. Any prolonged exposure to moisture and temperatures above 60°F will lead to sprouting. Dormancy in potato tubers, on the other hand, is hormonally controlled. After potato tubers go dormant, which is usually triggered when stems/leaves die back and tubers reach their final size, they may not sprout for weeks to months, even if exposed to excellent growing conditions.[6] This fact has vexed potato farmers in warm climates looking to grow multiple successive crops in a single year. As such, lots of research has gone into treatments that induce early sprouting in dormant seed potatoes.

Vernalization is another process to consider, even though it doesn't directly affect what happens in storage. It is the process that enables biennial plants to produce flowers after a period of exposure to cold temperatures. Many biennial crops go through a vernalization period during regular winter storage, but you won't see the results unless you grow the plants for a second season. This, of course, is important for seed producers, but most storage farmers don't care if crops are vernalized or not. Onion sets—small onion bulbs grown one season for planting the next—are one example of when growers use storage conditions that maintain dormancy but avoid vernalization. (For onions, these conditions are hot—above 85°F (29°C)—and dry.) Because most home and commercial growers keep onions in colder conditions, it's ill-advised to use small onions from storage as sets in the spring. Those onions will have gone through a vernalization period and will likely produce flowers when replanted.

Traits for Long Storage Life

While long-lasting dormancy is one important trait for storage varieties, there are other traits to consider when selecting both crops and cultivars for long-term storage. Size, shape, color, and taste are all important for marketing a crop, and yields are crucial to your bottom line. Two particularly important characteristics for storage crops are maturation time and disease resistance. To maintain quality in long-term winter storage, vegetables have to be both mature and healthy at harvest time.

It doesn't make sense to grow crops that won't mature before cold and darkness stunt their growth or damage them. Immature crops often have inferior taste, and small crops can be difficult to process and sell. Immature crops can also deteriorate quickly in storage. This is especially true for crops such as onions, winter squash, and pumpkins. Winter squash and pumpkins are non-climacteric fruits, meaning they don't continue to ripen after harvest. Once harvested, they won't develop the hard, waxy skins that help retain moisture and protect the inner flesh. Immature squash and pumpkins are also deficient in stored sugars and carotenoids that give their flesh its sweet flavor and bright orange coloring. It all amounts to fruits that dry out, decay, or succumb to wounds earlier than fully matured fruits. For most other storage crops, immaturity is a problem of size. Small crops will hurt your yields and efficiency, and they're the first to desiccate when storage conditions are too dry.

Unless a crop/variety will reliably mature within your growing season and with the season extension strategies available to you, it's not worth growing for storage. This might limit your options depending on your location. The number of frost-free days, soil temperatures, and daylength can all affect your ability to grow a crop/variety. Sweet potatoes, for example, need soil temperatures above 55°F (13°C) for at least 80 to 90 days, but preferably longer. Very short or cool growing seasons generally don't allow you to grow them profitably. Butternut-type (*C. moschata*) squashes are poor choices for interior Alaska and locations at similar latitudes not because of the short seasons—Hubbard-type squashes (*C. maxima*) consistently mature in time—but because they require nighttime darkness to set fruits. By the time darkness returns in late summer, there's not enough growing season left for the squash to mature before frosts arrive. Daylength also affects onions; if your days are too short for a variety, onions won't mature to form bulbs. Conversely, if your days are too long for a variety, the onions will bulb and mature too early, resulting in tiny onions and poor yields.

The health and integrity of crops at harvest time are crucial to long storage life. Promote healthy crops by supplying ideal growth conditions—healthy soil,

Cabbages at Root Cellar Farm in Toivola, Michigan, show signs of fungal rots after spending too much time in damp conditions under floating row cover.

stable watering, weed control, proper spacing, and so on—that encourage vigorous growth and discourage diseases and pests. Growers should also use practices such as crop rotation to avoid pathogen and pest buildup that cause damage and disease. That said, every variety will respond differently to the particular conditions on your farm. Some varieties might thrive while others struggle under the same set of conditions. It's not possible to know how a variety will respond to your farm's unique conditions without trialing (see "Variety Trials for Storage," page 10), but there are ways to improve your chances of success.

While it doesn't always make a difference, it's worth paying attention to where seeds are grown; if possible, choose seeds that were grown in a climate similar to your farm. It's also wise to choose seeds grown in similar cultural conditions to those on your farm. If you grow without synthetic fertilizers, for example, organically grown seeds are a smart choice because the genetic lines will be better adapted to those conditions. You should also pay attention to each variety's disease resistances and choose varieties that can thrive with your local plant pathogens. If you suspect that pathogens are affecting your crops, talk with other local growers and your local university extension to help diagnose the culprit(s). You will likely need to send in samples for a confident diagnosis because the visible symptoms of many plant diseases, nutrient deficiencies, and physical stresses manifest similarly.

Some pathogens prevent crops from yielding much at all. These are often those that attack the root systems or cause early leaf decay. Other pathogens can be less harmful to yields but affect a crop's quality and longevity in storage. These include gray mold (*Botrytis cinerea*), Phytophthora storage rot (*Phytophthora brassicae*), white mold (*Sclerotinia* spp.), and black mold (*Thielaviopsis basicola* and *Chalaropsis theilavioides*), to name a few.[7] Varieties that allow airflow between and under plants can also help mitigate certain disease pressures.

Lastly, you should look for varieties that align with your farm's systems and procedures. Do you plant with

tight spacings? If so, you may want more-compact varieties and those that allow for unrestricted airflow between plants. Do you use a mechanical harvester? Then your root crops should have tops strong enough for the harvester to lift from the ground without breaking. Think through the entire process—from planting to harvest to processing and storage—and choose varieties with traits that suit your systems. This might mean avoiding varieties that tend to break during harvest or washing, varieties that split or bolt easily, and varieties that tend to bruise or scrape easily. These types of damage can hurt your yields and lead to spoilage in storage. Again, the only way to know these things is through experience or trialing.

For some crops, it is easy to point to individual traits that affect storage life. For example, onions with less water content and more outer scales tend to last longer.[8] Often, though, it's difficult to isolate individual traits that consistently influence long-term storability. Take beets as an example. While traits like disease resistance, crown size, and cuticle (skin) thickness might logically affect storage life, there's not conclusive evidence linking storability to any single trait. Instead—as research professor and plant breeder Irwin Goldman from the University of Wisconsin told me—beet breeders select for long-term storability as a trait in itself. The same is true for most other storage crops; varieties that perform well in storage do so for multiple interconnected reasons. Plant breeders think of storability itself as a desirable characteristic and develop varieties that specifically perform well in storage.

When perusing the seed catalogs, storage farmers should look for varieties that stood out in storage trials. Seed companies sometimes comment on the storability of a given variety in addition to other characteristics. Take this example from the Johnny's Selected Seeds catalog, describing a new cabbage variety called Promise: "Best quality from storage. Fantastic right out of the field, or after long-term storage. Juicy, tender, sweet leaves—the best flavor in our spring storage cabbage taste tests. The dense heads are held high above the ground, which allows for good air circulation and reduced disease pressure. . . . Growers of Storage No. 4 will love this exciting new variety."[9] Based on the description, this is a variety worth trying for long-term storage. They note its long storage life, they comment on some characteristics that might contribute to that performance, and they even mention similarities to another popular storage variety to entice experienced growers who might be reluctant to try something new.

Unfortunately, seed catalogs often fail to describe how varieties perform in long-term storage. This doesn't necessarily mean a variety won't keep well in storage, but there are reasons to be cautious. If seed companies don't mention storage in the description,

it likely means the variety wasn't a standout performer in storage trials. It does not mean, however, that the variety won't keep for a time in storage. Yellowstone carrots, known for their deep yellow color and strong tops, are a good example of this. I've never seen a seed catalog mention anything about Yellowstone's performance in storage—probably because its color is its defining characteristic—but Yellowstone has consistently stored well on my farms.

In my experience, early-maturing varieties don't last as long in storage as main-season or late-maturing varieties. I suspect that early varieties have difficulty slowing down the fast metabolisms that allow them to grow so quickly. In storage, this could mean that they burn

Just because a seed company doesn't mention storage in a variety's description doesn't mean it won't do well. Yellowstone carrots (right) have consistently performed well in storage on my farms.

through energy reserves more quickly than slower-developing varieties, leading to early senescence and decay. This is my speculation, however, and I suggest that growers trial any variety they're interested in for long-term storage.

Variety Trials for Storage

There are many reasons to trial new varieties for storage. Let's say you're looking to add new colors, shapes, and sizes to your winter lineup. This is a great way to increase diversity to your CSA or differentiate yourself

Crew member Anja records harvest weights of winter squash varieties at Offbeet Farm. *Photo courtesy of Phil Knapp.*

from other vendors at market. Alternatively, you could be trying out a new storage variety similar to one you're already growing. Trialing allows you to see how varieties perform before relying on them too much. Trials can also help you fine-tune your business by identifying varieties, even those you're already growing, that best fit your needs. Some universities and cooperative extensions publish variety trial results freely online. Although these results can be insightful, conducting your own trials can reveal how individual cultivars behave in the unique conditions on your farm.

There are trade-offs between the time and effort you put into planning and executing a trial and the quality and accuracy of the results. Thus, you need to decide how much you're willing to expend at the outset. At one extreme, you could grow a few new plants and make some simple observations throughout the growing season and in storage. At the other extreme, you could design a full-blown experiment with randomized and replicated plots, a set of predetermined measurements, and statistical analyses of the data. I advocate for a middle road. As a working farmer, you probably don't need research-quality results, but some basic data collection will help you make better decisions than observations alone.

Your needs and interests will determine what traits you track in a trial. Conventional trials often focus on yields, size, aesthetics, disease/pest resistance, bolt resistance (if applicable), and taste, whereas storage trials track longevity and quality in storage. That said, your trial objectives don't have to be mutually exclusive. Before designing a trial, think about the ways you process and sell your crops and what information would be useful to know. You may be interested in which varieties last longest in storage and offer the best yields. You might want to find varieties with the best disease resistance or know when to sell or eat those varieties before suffering heavy losses in storage. You don't need to track everything. Variety trials should help you make informed decisions to improve your farm and business rather than being unnecessary burdens.

For most farmers, the most difficult part of variety trialing is the data collection. Plan ahead and think of ways to efficiently collect information alongside your current workflows. This is one argument for keeping regular planting, harvest, and sales records on the farm. If you're already accustomed to collecting information, it's easier to collect the data necessary for good variety trials. Choose a method of collecting information you'll actually use. This might mean printing your data collection sheets ahead of time and leaving a clipboard in a convenient and visible location. It might mean designing a method of digital data collection on your phone or other electronic device, especially if that device is regularly in your pocket. I personally like Google Forms, which are free, fast to create, and completely customizable. The data you enter automatically populates in spreadsheets you can access later, but there are other apps and programs that do the same thing.

Another important part of any trial is ensuring that the varieties you're testing get similar treatment throughout the growing season, harvest, processing, and storage. The goal is to compare the varieties themselves and not the different conditions that occur on your farm. The varieties you're trialing should have similar planting dates (ideally the same day), soil and watering conditions, harvesting dates and procedures, washing procedures, and storage environments. If you treat varieties differently, their yields, quality, and longevity in storage may be affected. In practice, though, it's difficult to treat every variety exactly the same. Just do your best to limit differences where you can, and those small differences shouldn't overly affect your results.

Sometimes taking measurements of smaller samples is more practical than measuring an entire crop. This, of course, depends on the amount you plant and what you're measuring. Do you have half an acre of purple carrots to test, or do you have 20 row-feet? It might be practical to measure yields from a half-acre of carrots, but it's not practical to measure spoilage rates in 15,000 pounds of stored roots each month. Sampling allows you to efficiently measure and represent large groups. It's important, though, that samples are both unbiased and representative. If you select your samples primarily from vigorous and healthy parts of a planting while ignoring diseased or underperforming areas, the results of your trials will be skewed. Instead, collect samples as randomly as possible.

How large should your samples be? Some recommendations suggest at least 50 plants for traditional root vegetables, at least 10 plants for tubers, and at least 30 plants for nonroot brassicas.[10] For small crops such as garlic and sunchokes, I suggest samples of at least 10 pounds (4.5 kg). For large crops such as cabbages and winter squashes, samples might need to be upwards of 100 pounds (45 kg). For everything in between, samples of around 50 pounds (23 kg) should suffice. The larger the samples relative to the whole, the more accurate your results will be. That said, if you only have time to measure a few plants of each variety you're testing, some data is always better than none.

Below, I describe a few specific measurements I find valuable for storage variety trials, and I've included tips for sampling, data collection, and calculations. There are certainly more measurements you could take, but I find these particularly useful for planning annual production, sales, and marketing.

Yields

I think it's vitally important to know the yields from each variety I grow. If weighing yields from an entire planting is too onerous a task, you can weigh yields from smaller sample plots, but you'll need to know the area sampled for calculations. (Areas in this context can be any planting space—acres, row-feet, beds, and so on—compatible with your farm's layout and practical for planning purposes.) You can include or exclude culls in this measurement depending on your typical culling methods (see below). I measure yields while putting crops into long-term storage, noting the total weights of each variety separately. Later, I check my planting records to see the area of each variety planted. The calculation is the harvested weight divided by the area planted (or sampled).

Culls and Grades

Knowing the amount lost through culling helps you determine a variety's actual value compared to its potential. The same goes for the amounts sorted into different grades. Culls are pieces you reject outright because of disease or defects, whereas grades are salable categories used for marketing a crop (see "Grading," page 42). You can calculate culls and grades as percentages of harvested weight. These measurements don't depend on the area planted/harvested, and I suggest measuring from samples. The overall sample weights won't matter too much, although ideally, you should sample similar amounts from different varieties of a given crop. For each sample, measure the total weight, the weight of culls, and the weights assigned to each grade used on the farm. To calculate each category's weight as a percentage of the whole, divide cull/grade weight by the sample weight, and multiply by 100 percent.

Loss Through Time in Storage

This is one of the most useful pieces of information for marketing crops on a winter storage farm. Through trialing, your goal is to determine the proportion of a variety lost to desiccation, sprouting, and rot at different times in storage and to identify when losses become too high to justify keeping that crop any longer. First, you'll need to decide what you consider to be unsalable in your specific markets. For example, how much sprouting is too much? Are you trimming off spoiled portions? You'll also need to devise a regular sampling schedule. The frequency is up to you. I like to sample at one-month intervals because it matches how I plan my wintertime CSA and wholesale sales.

Measuring losses from thousands of pounds repeatedly won't be practical, so you'll need to devise a sampling method. One method is to take measurements periodically while packing orders. To do this, weigh a sample that you will process for sales. Set aside and weigh trimmings and other unsalable pieces and calculate their percentage of the total weight sampled. Another method is to set aside samples in separate containers to measure periodically. Weigh

Variety Trial Datasheet

Year: _____   Crop: _____   Variety: _____

Planting Date: _____   Harvest Date: _____   Location: _____

Yields

Sample Weight: _____   Yield = Weight/Area

Sampled Area: _____

Culls and Grades

Sample Weight: _____

Cull Weight: _____   Cull %: _____

Grade 1 Weight: _____   Grade 1 %: _____

Grade 2 Weight: _____   Grade 2 %: _____

Grade 3 Weight: _____   Grade 3 %: _____

Loss in Storage

Date 1: _____   Sample Weight: _____   Loss %: _____

Weeks in Storage: _____   Loss Weight: _____

Taste (1–5): _____   Comments: _____

Date 2: _____   Sample Weight: _____   Loss %: _____

Weeks in Storage: _____   Loss Weight: _____

Taste (1–5): _____   Comments: _____

Date 3: _____   Sample Weight: _____   Loss %: _____

Weeks in Storage: _____   Loss Weight: _____

Taste (1–5): _____   Comments: _____

Date 4: _____   Sample Weight: _____   Loss %: _____

Weeks in Storage: _____   Loss Weight: _____

Taste (1–5): _____   Comments: _____

Date 5: _____   Sample Weight: _____   Loss %: _____

Weeks in Storage: _____   Loss Weight: _____

Taste (1–5): _____   Comments: _____

Date 6: _____   Sample Weight: _____   Loss %: _____

Weeks in Storage: _____   Loss Weight: _____

Taste (1–5): _____   Comments: _____

An example datasheet for collecting storage trial information for a specific variety. I highly recommend finding a data collection method you'll actually use amid the occasional chaos of harvest time.

each sample at the beginning of storage, then periodically remove and weigh the unsalable portions. While the first method allows you to sample from within your regular storage containers, you won't be able to determine weight lost through desiccation in storage. (Besides spoilage and sprouting, water loss will be one of the largest sources of lost income in storage.) If you sell out of a variety early, you'll limit the breadth of your data unless you hold a portion back for later sampling. The second method allows you to account for water loss in storage, but I'd advise against it if the sample containers are different from those you use for the main crop. (Ideally, you want sample containers to match the regular containers you use for that crop.)

One final method, which I personally like, is to set premade samples in storage alongside the main crop, using plastic mesh bags to keep samples separate. This method requires forethought, but it also means the samples will experience the same conditions as the main crop and you'll be able to track water loss in storage. Again, you need to measure each sample's total weight at the beginning of storage. At the time of each measurement, remove and weigh all unsalable portions. Calculate the weight of unsalable portions as a percentage of the sample's weight at the beginning of storage.

Taste and Aesthetics

Although these are subjective measures, it can be useful to know how a variety's looks and tastes change as it ages in storage. Subjective traits are often measured according to a numbered scale with predefined categories. Here's an example of a 1 to 5 scale for taste: 1) Completely unpalatable; 2) Tastes bad but is edible; 3) Neither good nor bad; 4) Tastes good but is not the best example of this crop; 5) Bliss. Here's an example scale for aesthetics: 1) Rotten or unsalable; 2) Ugly, but I'd eat it; 3) Looks OK; 4) Looks nice but with minor blemishing; 5) Looks freshly harvested. Of course, you can define your own categories to suit your needs. I suggest taking these measurements alongside those for losses in storage described above.

One measurement I've intentionally omitted here is overall losses in storage. This can be a vital data point for crop planning and could, in theory, be part of a variety trial, but I think it's more relevant as part of your regular record keeping (see "Record Keeping," page 70). That's because overall losses in storage depend not only on the proportions lost over time but also on sales rates. Since you'll be selling or eating portions of your crops throughout the storage season, the total amounts lost in storage will differ depending on how much of a given crop remains. For example, if you expect to lose 10 percent of your remaining carrots in March, you'll lose more if 5,000 pounds remain than if only 1,000 pounds are left. Without knowing how quickly you'll sell a variety, you can't predict the overall storage losses.

Chapter 2

Harvest

As a storage farmer, I love harvest time. It's when the culminated efforts of a summer spent planting and tending come to fruition. Thoughts of changing leaves and ripening pumpkins fill many people with warm nostalgia for a distant agrarian past. But for the storage farmers out there, the warm feelings can be accompanied by the anxiety of an all-too-real agrarian present. While I savor working in the crisp fall sunshine, I'm sometimes distracted, worrying about all the potential problems that arise when bringing crops from the field into storage. Harvesting and processing for long-term storage is more consequential than it is for fresh marketing since small problems can balloon over time. A crop that normally stores for six months or more might last only a few weeks if badly damaged—by frost, for example. Likewise, a hard bump during washing might lead to mold growth that takes out not only

Despite the sometimes-frantic pace, harvest days are some of my favorites of the entire year. *Photo courtesy of Phil Knapp.*

that piece, but all those around it in a few months' time. Mistakes during harvest and processing can ruin months of hard work, leading to disappointment and decay in storage.

Thankfully, successful long-term storage is an achievable goal; it just takes care and knowledge while bringing crops from the field into the storage room. The first step of that process is harvesting, and I'm a strong advocate for thoroughly planning out the harvest season. Planning removes much of the guesswork and gives you small steps for tracking your progress. The harvest season doesn't need to be stressful. While some things will always be out of your control—the weather, for example—planning and experience can guide you through the unexpected and toward smooth harvests and lasting storage.

Harvest Order

If we had all the time in the world, the harvest would be an easy task. But there's one immutable fact that puts a timer on the whole operation. As George R. R. Martin helpfully reminds us, winter is always coming.

We need to ensure that our crops are fully mature and that we have enough time to harvest and process them before winter inevitably arrives. When farming for storage, we also need to get crops from the field and into storage in the best condition possible. Ideally, the crops have reached the desired size, are free from diseases and damage, and have been exposed to the right sorts of temperatures along the way.

Planning an order for harvest is a good place to start. A general harvest order comes naturally in many climates, and most storage farmers across the northern United States and Canada follow similar harvest patterns. (Many farmers in the southern United States also use harvest orders similar to their northern neighbors, albeit more drawn out.) Regardless of where you live, the order of harvest will largely be determined by crop maturity and the way each crop reacts to changing patterns of weather and daylight. It's common that multiple storage crops become ready for harvest simultaneously, which calls for some decision-making. While some crops are content to wait patiently in the field, others deteriorate if left too

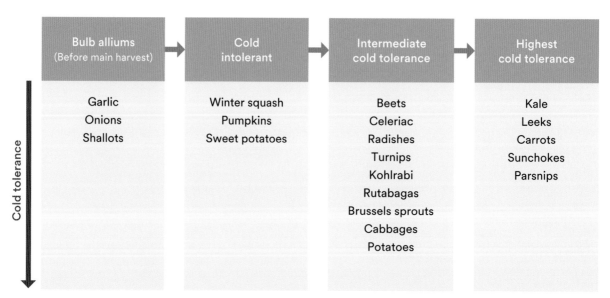

The harvest order for storage crops is based on when given crops mature and how tolerant each is to cold temperatures. Here, crops are grouped based on time of maturity (for alliums) and tolerance to cold, moving from left to right. Within each group, crops tolerate varying levels of cold, with more sensitive crops toward the top and hardier crops toward the bottom of each list.

long. Diseases, pests, availability of labor, and your ability to cool crops postharvest are other factors that will influence your decisions. While I make a harvest plan each year, I also watch crops and the weather closely to triage if necessary.

Many storage farmers begin their harvest with allium crops, which are special in that their maturation is driven largely by daylength. Compared to other storage crops, garlic and onions mature relatively early, and the signs are unambiguous. They're also both easily damaged if left overlong in the field past maturity; garlic cloves tend to grow too large and split, while onions often develop fungal infections when left unharvested during wet weather.

Next, it's common to harvest the most cold-sensitive crops, including pumpkins, winter squash, and sweet potatoes. Winter squash and pumpkins are easily damaged from frost exposure, with damage manifesting as discoloration and small pockmarks on the skin. Winter squash and pumpkins also accumulate chilling injury below 50°F (10°C) that leads to early deterioration in storage. Just like winter squash, sweet potatoes accumulate chilling injury that shortens their storage potential. Although sweet potato vines and foliage die from any frost exposure, the soil temperature is more relevant to chilling injury than air temperature. Exposure to soil temperatures below 55°F (13°C) begins the process of degradation.

As fall progresses, storage farmers turn their focus to crops that can tolerate mild exposure to cold and frosts but are easily damaged by hard frosts. Common storage crops in this group are the brassicas—cabbage, kohlrabi, rutabagas, turnips, and winter radishes (such as daikon)—as well as beets, celeriac, and potatoes. I would also include radicchio in this group, though I'm on the fence about radicchio as a true storage crop because of its short storage life (for more on radicchio, see "Boldly Grown Farm," page 227). Most of these intermediate crops have their salable portions either partially or fully aboveground, and the main concern is freezing damage. Allowing any parts destined for storage to freeze, even on the

Leaves protect the beetroots below from damage during light frosts at Offbeet Farm.

surface, is a problem. Common gardening advice says they can handle exposure to temperatures as low as 28°F (−2°C). Although it feels comforting to have a specific boundary, there's more nuance to cold exposure than just the temperature (see "Cold Exposure," page 22). Despite the fact that potato tubers are entirely underground, I include potatoes in this group because of their sensitivity to frozen soil. As opposed to the hardiest group of crops, you will have big problems if potato tubers come into contact with frozen soil. Even if a majority of the tubers are fine, there will be pieces with freeze damage manifesting as softened, watery flesh. Any damaged pieces you miss during sorting and put into storage will liquify into stinky brown goo that ruins most pieces around it.

Within this intermediate group, you may have reasons to harvest crops in a specific order. Beets are particularly sensitive to cold and should be one of the first out with hard frosts approaching. Even light

Parsnips are one of the last crops harvested on Offbeet Farm because they're among the hardiest to cold.

freezing on the root will lead to surface mold growth in storage, and their leaves aren't as protective as some of the others in this group. Cabbages and radicchio offer some forgiveness since you can peel away freeze damage, but for the others, you may be out of luck. Pest pressures such as cabbage root maggots might push you to harvest brassicas as soon as possible; alternatively, you may find yourself trying to prevent damage from overmaturation such as splitting in cabbages and kohlrabis or bolting in daikon radishes. I've also found that some of these crops are more tolerant of harvesting in warm weather than others. In my experience, cabbages, kohlrabis, turnips, and potatoes are more forgiving than the others if you're struggling to cool them quickly postharvest.

The final group of crops to harvest are the hardiest to cold exposure. These include carrots, parsnips, sunchokes, leeks, and kale (if you're counting it as a storage crop). With the exception of kale, the salable portions of these crops are almost entirely belowground and can generally tolerate some exposure to frozen soil. In fact, the flavor of these crops improves with frost exposure. That said, you do not want any portion of the roots, tubers, or (for leeks) the blanched portions of tightly bound leaves to freeze. Kale leaves can freeze solid and return to fair condition but with some damage to their fleshy midribs. The aboveground leaves on leeks act similarly when frozen, often developing odd textures and performing poorly in storage. Carrot crowns that poke up above the soil surface can easily freeze. The damaged tissues usually appear cracked and spongy and will be among the first to spoil in storage. Farms with heavy soils or in areas with wet weather in late fall often choose to harvest their deep-rooted crops relatively early. Mechanical harvesters struggle in wet soils, and clumped clay makes digging and washing these veggies difficult. In my experience, all of these crops appreciate being cooled as quickly as possible postharvest.

Inevitably, a stint of unexpected cold weather will leave you in a pinch to harvest cold-sensitive crops. Covering crops with either row cover or mulch can buy

you some time. Row cover can protect against light frosts and, depending on thickness, gain you a few degrees. Silage or standard woven tarps and straw can buy you even more protection and prevent the ground from freezing for a time. Sometimes, though, you'll just need to burn the midnight oil to get crops in safely.

Scheduling the Harvest

Once you've established an approximate harvest order, it's a good idea to plan out, or at least think about, a schedule. This is especially true if you have a lot to harvest or have other commitments like deliveries or markets during harvest time. When scheduling my harvest, I use two main pieces of information to guide my decision-making for some crops: latest harvest (or deadline) dates and an estimate for how long it will take to harvest each crop. Unfortunately, it takes experience growing vegetables in your location to know this information with certainty. Even with experience, I find past climate data from my nearest weather station helpful while making these decisions.

Finding Local Climate History Data

In the United States, the National Weather Service maintains freely accessible databases of past climate data for its primary weather stations. (In Canada, Environment and Climate Change Canada collects and disseminates this type of information.) The National Weather Service website has a special page for Past Weather. There, you can find past data from regional forecast offices organized in different ways. One option is to look at daily/monthly normals. If you select daily normals, you should see tables showing average high, low, and mean temperatures in that location for every day of the year.[1]

You can use historical mean daily temperatures as approximate guides for harvest scheduling. (The mean daily temperature is the average temperature across a 24-hour period.) For crops that experience chilling injury, you should plan to harvest on or before the date when the mean daily temperature reaches the threshold for injury. Frost damage often spells disaster, but chilling injury, while less serious, can meaningfully shorten the storage lives of affected crops. On my first farm near Houghton, Michigan, that date for winter squash (which experience chilling injury below 50°F, or 10°C) was October 1. At my current farm in Fairbanks, Alaska, the date is closer to September 7. Both dates are about a week before the typical first frosts, so they're opportune times to remove squash from danger. It's also important to get even the most cold-tolerant crops out before the ground freezes, which begins when mean daily temperatures drop below freezing. At the Michigan farm, this date averaged around November 15, whereas in interior Alaska, it's October 10. These dates are the deadlines for completing the harvest, which should preferably be done a week or more beforehand. I cannot emphasize enough that chiseling carrots or parsnips out of frozen ground is anything but fun.

I use the examples of winter squash and highly cold-tolerant crops like carrots and parsnips because these are the crops that I, and many storage farmers across the United States and Canada, use to frame the main harvest season, when we're bringing in the majority of our storage crops. Farmers growing sweet potatoes will also need to assess soil temperatures, which typically lag a few days behind air temperatures during seasonal changes. At Offbeet Farm in Fairbanks, the bulk of my harvest occurs roughly between September 4 and October 6. I start a little early (before daily means drop below 50°F) to buy extra time toward the end of the season when cold weather encroaches, and I'm able to do this because some crops are ready by early September. Also note that for me, garlic and onions don't fall into these considerations.

Near Fairbanks, they're mature and ready to harvest in August, before the main harvest window.

For your own farm, come up with start and end dates for the main harvest season. Again, use your own experiences, the experience of others, and historical weather patterns as your guides. If you're not growing winter squash or sweet potatoes, then find an approximate start date for your least cold-tolerant crop. This might be the day when the average minimum temperature drops below 28°F (−2°C) for the intermediate crops, for example. It's also a good idea to give yourself a week or more of buffer. Each year's weather patterns are different, and it's smart to plan for year-to-year variations.

After identifying a possible harvest window, you'll need to estimate how much time the harvest will take to complete. This is difficult to generalize because it's so specific to each farm, farmer, and crop. On Offbeet Farm, the crew and I can harvest a block of ten 100-foot cabbage beds in a little over a day, whereas the same area of carrots might take us four days to complete. The tools and machines available to you will make a huge difference. On my farm, two people can hand-harvest about 800 to 1,000 pounds of carrots in an eight-hour day, but this includes digging them with a harvest fork and carting them over to the washing station by hand. If I could undercut the beds prior to pulling carrots, the harvest rate might double (see "Tractor-Run Harvesters," page 25). With a dedicated carrot harvester, harvesting 1,000 pounds might take only a couple person-hours. If you're unsure about your harvest speed, my advice is to ask around and to take good notes during or immediately after harvest season. Ask a fellow farmer to estimate how long harvesting specific crops might take with the tools available to you. I also advocate for conservative planning. If your estimates are wrong, you'd rather have more time than not enough.

Using your harvest time estimates for each crop, you can estimate how long the main harvest will take. Again, exclude the time it takes to harvest any crops outside of the main harvest window. Now look back at the start and stop dates you came up with and

Table 2.1. Harvest Planning Example

Crops in Order of Harvest	Harvest Equipment	Number of Beds	Harvest Rate	Days
Winter squash	Hand	20	12 beds/day	1⅔
Beets	Hand	18	6 beds/day	3
Kohlrabi	Hand	12	6 beds/day	2
Turnips	Hand	12	6 beds/day	2
Cabbages	Hand	30	15 beds/day	2
Potatoes	1-row digger	30	10 beds/day	3
Carrots	Undercutter-Hand	30	5 beds/day	6
Parsnips	Undercutter-Hand	20	6 beds/day	3⅓
Total		**172**		**23**

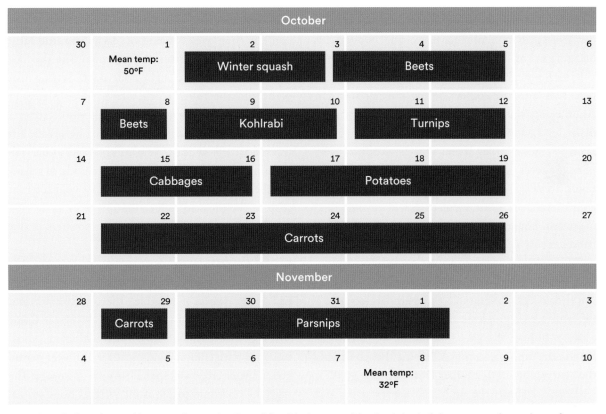

Based on their estimated harvest times, the Grand Rapids farm could schedule their harvest as shown here, from the least cold-tolerant crop to the most. They should start harvesting squash before average daily mean temps drop below 50°F and finish their entire harvest before average daily mean temps drop below 32°F—preferably at least a week sooner.

Broadfork + Hand Harvest	Undercutter + Hand Harvest	Mechanical Harvester
100–200 pounds / person hour	200–300 pounds / person hour	1,000–2,000 pounds / person hour

Harvest rates vary depending on the crop and the tools available to you. These harvest rates were made with carrots in mind, but they are applicable to other root crops as well.

compare that range to the estimated harvest time. How do they compare? Is there more time than you need to complete the harvest? If so, that's great! There are usually plenty of other tasks to keep you and your crew busy, such as onion cleaning or fall cleanup. If not, what changes do you need to make in order to finish the harvest on time? One option is to start earlier. You'll gain more time for harvesting, but you risk bringing in smaller or immature crops unless you're able to adjust planting dates. You'll also bring more field heat into the coolers during warmer weather. Another option is to hire more labor. Many hands make for light—and fast—work. Of course, you'll need to train and pay them. One final option is to purchase tools or adjust systems to make the harvest faster. Tools and machinery are great ways to increase your harvest speed, but not all tools will fit your farming system. For example, I cannot use many mechanical harvesters on my 30-inch beds.

While these general principles guide harvest planning at Offbeet Farm, other farms harvest differently to suit their needs. For example, Southern farmers have a different set of problems that drives their decision-making. Instead of cold temperatures, it's often heat around the time of planting or harvest that constrains their harvest timing (see "Open Door Farm," page 197).

Table 2.1 shows the estimated harvest rates and times for an imaginary farm located near Grand Rapids, Minnesota, where mean daily temperatures drop below 50°F on October 1 and below 32°F on November 8. This fictional farm has 2 acres of storage crops, and their field setup consists of 100-foot beds that are 5-feet wide on center. The harvest rates are based on a harvest crew of two people working for eight hours each day. The crew has access to a tractor along with an undercutter and a one-row potato digger. After making some conservative estimates for the harvest rates (given their available equipment) and the time they'll spend on each crop, the farm predicts the two-person crew can complete the harvest in twenty-three eight-hour days. Setting this plan onto the calendar, they find there's plenty of time to complete the harvest between the October 1 to November 8 window. They'll even have time for weekends! Or at least time to complete other fall cleanup, with a buffer to boot. They decide to schedule the harvest in order from the least cold-tolerant crop to the most to reduce the risk of an early cold snap damaging crops.

Cold Exposure

There are strong arguments for harvesting on the later end of the season if possible. The cold-tolerant and cold-hardy storage crops often develop better flavor after exposure to cold weather. By harvesting later, you'll run your refrigeration equipment for a shorter period because your crops will spend less time in storage. The vegetables become cooler as the weather does, so you will reduce the amount of field heat brought into storage and the time needed to cool crops. Most farmers agree that storage quality improves the faster the crops cool down to storage temperatures after harvest.

Judging the cold-hardiness of certain crops can be tricky, though. Common advice sets boundaries at 28°F (−2°C) for the intermediate crops and 24°F (−4°C) for the hardiest crops, but these numbers don't tell the whole story. When thinking about a crop's tolerance to cold, you should consider both temperature and duration of exposure, as well as the current mean daily temperatures. Generalized advice can be useful, but it doesn't always translate across geographic areas and different weather scenarios.

Let's use winter squash as an example. Imagine one farm growing winter squash in the high desert of New Mexico, where daytime temperatures are relatively warm—say 65°F (18°C)—and nighttime temperatures are dropping just below freezing for the first time in the season. Squash fruits hold onto much more heat energy than the leaves, so they respond differently to brief periods of below-freezing temperatures. Because the thin leaves can't hold onto much heat, they will stay relatively close to the air temperature. When the air briefly drops below freezing, water in the leaves freezes and the tissues are damaged irreparably, quickly discoloring and drying the next day. Meanwhile, the fruits can hold onto more heat from the day and, even without a protective canopy of leaves, can sometimes tolerate brief stints of below-freezing air without freezing themselves. The fruits might not even sustain chilling injury if the flesh manages to stay above 50°F (10°C).

Now imagine a different farm growing winter squash in Alaska. There, it's common for frosts to appear relatively late in locations elevated above valley floors, but daily highs can be chilly even if frosts haven't arrived. Let's say the high temperature the day of the first frost is 42°F (6°C). (For the sake of example, let's ignore chilling injury to the squash.) Compared to squash at the high-desert farm, these squash won't have as much heat energy from the day to carry them through the night. If temperatures briefly drop to 31°F (−1°C), there's a higher chance the fruit might lightly freeze near the surface.

A Winter Luxury pumpkin gets some warmth from the sun and attention from a bumblebee.

This example illustrates that low temperatures are only one piece of the puzzle. It's important to note not only minimum temperatures, but also how long the low temps will last, as well as high temperatures and daily means. All the crops commonly considered hardy down to 28°F or 24°F are only safe for a brief time, but more so if daily means are warm. If temperatures remain below freezing for more than a few hours, these crops are in trouble regardless of daily means. As the weather changes, the temps of large watery veggies like winter squash will closely match mean daily air temps, whereas soil temperatures often lag a few days behind. When daily means reach the point at which crops are damaged, it's time to harvest them.

Harvest Tools

While there may be some overlap with other vegetable farms, farms that focus on storage crops usually have specialized tools and equipment to make the harvest easier and faster. Nearly all storage crops are bulky and heavy. Many storage crops grow at least partially underground and must be dug out. Most tools and equipment for harvest either help you dig up crops or move them from place to place more efficiently. Some tools are common to most farms, whereas others are more specific to either small- or large-scale farms.

As storage farms grow and develop, many are on the search for an equipment sweet spot. The purpose of machinery is almost always to make tasks faster, more efficient, and more ergonomic to workers. Machines come with costs, some obvious and some not, and any farmer should estimate cost savings when considering a new piece of machinery. How much does it cost in labor, fuel, time, and so on to do a task currently, and how will that change with the new equipment? Are those savings worth the costs of the equipment? How much labor is available, and would you rather hire workers or use a faster machine? Will the equipment prevent hidden costs, such as future medical bills or loss of physical ability due to injury or overuse? Machines and equipment should match not only your budget and scale, but also your goals and values.

Cutting tools are found on most every farm, regardless of size. These are simple knives and pruners for cutting and trimming crops in the field. Knife styles are a personal choice. For example, some farmers prefer lettuce- or broccoli-style harvest knives—which have sharpened and squared-off or rounded tips—for cabbages. These allow workers to cut through stalks with a pushing motion that utilizes the shoulder rather than the wrist. Many farmers also prefer pruning shears over knives for harvesting winter squash and pumpkins. Even if stems aren't corky, there's no telling what you might hit after cutting through a resistant squash stalk (another squash, irrigation lines, or worse). With both knives and pruners, it's important to keep them sharp, and I usually carry a sharpener in my back pocket. Sharp knives and pruners are safer to use than dull ones, and they make cleaner cuts less likely to attract pathogens in storage. It's also a good idea to attach bright colors to these small and easily-lost tools! Whether that means spray paint, tape, or flagging, you'll thank yourself for doing this when a tool suddenly disappears.

There are several tools that help loosen soil so that belowground crops are easier to pull by hand later. At the hand scale, there are both digging forks and broadforks. Digging forks, also called spading forks,

are great tools for many tasks, from loosening carrots to digging potatoes. They're easy to use and inexpensive, but they're relatively slow. Broadforks are a faster hand-scale option. They're usually two to three times wider than traditional digging forks, but they often cost considerably more. Although you can use regular broadforks for harvesting, the harvest-specific broadforks are best. Their tines are closer together—to reduce the risk of tines slipping through the soil and damaging crops—and narrower, so they're lighter and easier to push into the ground. The go-to tools on larger, tractor-scale farms are undercutter bars, also called bed lifters. These are simple, relatively inexpensive implements mounted to the three-point hitch on full-sized tractors. The "bar" is pulled through the soil

A potpourri of harvest knives at Offbeet Farm. The ones with bright orange markings—either paint or flagging—are much more likely to be found if lost in the field. *Photo courtesy of Phil Knapp.*

at depths of 12 to 16 inches to lift and loosen the soil above (hence the name). Many farms without more specialized harvesting equipment run undercutter bars beneath their deep-rooted crops to make hand-pulling easier and faster. I've also heard of farms using subsoilers to loosen soil around crops like garlic, but undercutter bars are far more common. Unfortunately, two-wheeled tractors don't have enough weight or traction to pull undercutter bars through the soil. Many have tried it, and many have failed.

Tractor-Run Harvesters

Both for two-wheeled and full-sized tractors, there are more specific harvesting implements designed for single or multiple crops, though there are fewer

The potato harvest at Open Door Farm in Cedar Grove, North Carolina, assisted by a single-row potato digger. *Photo courtesy of Jillian Mickens.*

options for two-wheeled tractors. Both BCS and Aldo Biagioli sell relatively inexpensive root-digging plows for two-wheeled tractors, but the general consensus is that these apply mainly to potatoes. You'll need to add weight to the two-wheeled tractors—usually wheel weights—and the diggers go only as deep as the soil has recently been worked. Crops such as sweet potatoes, carrots, and daikon radishes are often rooted too deeply and are damaged by these diggers, whereas more shallowly rooted crops such as beets, turnips, and onions don't require enough soil-loosening to warrant such a tool. These implements might be useful for garlic and sunchokes if driven alongside the rows. There are also PTO-powered potato diggers for walk-behind tractors, but they, too, have a depth limitation. The manufacturers suggest that you plant potatoes shallowly and rely on hilling to effectively use these harvesters.

Potato diggers for full-sized tractors are some of the most popular harvesting implements out there. They're relatively inexpensive and simple to run and repair, often lacking any hydraulic components. Storage farmers use them for potatoes, of course, but they're also effective at harvesting sweet potatoes and sunchokes. Most farmers I've talked to use Spedo-brand diggers. The simplest and cheapest model is a one-row shaker, which sieves out soil and lifts tubers to the surface. It's similar to the walk-behind digger but can dig up to 12 inches into unhilled soil. That said, both weeds and wet soils tend to thwart this style of harvester. As one farmer I spoke with reported, the vibrating diggers struggle to separate soil clumps and end up slicing through sections of wet or weedy soil without actually sifting out the spuds. Spedo (and other companies) also sell chain-style diggers, which use chain conveyors to remove soil from around tubers before depositing them on the soil surface. Chain diggers are more expensive but can sometimes harvest multiple rows at once and work in a wider variety of soil conditions than shaker-style diggers. There are also larger chain-style harvesters that

Carrots dangle from the harvest belts of a Scott Viner on Food Farm in Wrenshall, Minnesota. *Photo courtesy of Janaki Fisher-Merritt.*

incorporate features such as sorting areas for workers and the ability to carry and fill bulk pallet bins.

Scott Viner Harvesters

The FMC Scott Viner (commonly called the Scott Viner) is by far the most popular root harvester among small-scale storage farmers. Every storage farmer I talked to who mechanically harvested their carrots, beets, and other roots either had used or was currently using a Scott Viner. Originally designed for beet harvesting in the 1930s, Scott Viners are used today for harvesting carrots, beets, radishes, parsnips, turnips, and even rutabagas, depending on modifications to the machine. Their popularity is due in part to their functionality, but also to their price. Because they're no longer manufactured and most are old, they're relatively cheap as far as implements go. The newest machine I could find for sale was made in the 1970s. That said, they're relatively simple machines, and if a farmer is willing to replace hydraulic lines, grease or replace bearings, and resurface rusty metal, older Scott Viners do their jobs admirably. They're also versatile and can be modified to work within a specific farm's systems.

Scott Viners work by running a digging shoe beneath a row of roots—let's say carrots for example. At the same time, a set of rubber belts pinch carrot tops just above the crowns and lift them from the ground. The belts run backward and up, the carrots dangling by their tops, toward a knife that cuts the tops and drops the carrots onto a chain conveyor. The tops are shot out the back of the machine while the roots are conveyed away, the soil bouncing through the chains. Some farmers drive a wagon alongside the harvester for the carrots to drop into containers, while others modify the conveyor to spit carrots out behind the machine.

Scott Viners and similar harvesters can turbocharge the rate of harvest for deep-rooted crops like carrots. Several farmers reported that one fast worker can harvest between 200 and 250 pounds of carrots per hour on a clean bed that's been loosened by an undercutter bar. With a well-functioning Scott Viner, the typical rate increases to 2,500 to 3,000 pounds per hour. However, running the Scott Viner might require three to four people: one to drive the tractor, one to operate the Scott Viner, one to funnel carrots into bins, and one to pull a wagon alongside the machine. Scott Viners also require relatively clean and dry fields. Weeds tend to block up the topping mechanism and can make it difficult to see and align the machine to the row. The machine also has trouble staying aligned and pulling roots from heavy, wet soils. Many farmers running Scott Viners or similar machines shift their harvest dates to increase their chances for dry weather.

There are newer machines similar to FMC Scott Viners that require fewer people to operate. One popular example is from the Canadian company Univerco. Their smallest root harvester can be operated by a single person who drives the tractor and operates the machine while the implement fills bulk pallet bins. These machines can also offload full bins with a hydraulic platform lift. Although such machines are more expensive than Scott Viners, they typically run with fewer mechanical issues and have lower labor costs to operate.

Materials Handling at Harvest

Harvest is usually the first point in the storage process when materials handling becomes an issue. Storage crops are generally bulkier, heavier, and less valuable by weight than their fresh-marketed cousins. To be a successful storage farmer, you need to be able to lift and move heavy objects from place to place quickly and safely.

I like to think of the process of moving crops out of the field and into storage like a river system. Furthest upstream are the small drainages and tributaries collecting water from small swales and ravines. These are like veggies passing through the hands of workers as they harvest small sections of the field. As water moves downhill, the small streams join together to form larger rivers. These are like veggies from individual sections of the field aggregated together for processing steps such as washing or curing. Eventually, more and more tributaries join to form large and powerful rivers that drain entire watersheds. After processing, we place veggies inside our largest bulk containers and funnel them inside our storage spaces where they remain for weeks to months. What starts as a trickle becomes a mighty river of vegetables collected in one place.

There are two main pieces to the puzzle of materials handling at harvest (and in general): the containers used to hold crops and the equipment used to move them. Containers and handling equipment are linked; the containers you use influence your options for handling equipment, and vice versa. This relationship also extends to the equipment and methods used for processing steps before storage, but we'll cover that in the next chapter.

Let's start with options for materials-handling equipment. At the hand scale are things like wheelbarrows and carts. At Offbeet Farm, we mostly use a modified garden cart to transport veggies from the field to the wash station. This allows us to move about 200 pounds at a time with minimal effort, so long as loads are well balanced. Carts and wheelbarrows usually have limitations for both the weight and height (when stacking multiple containers) because the user tips them at an angle in addition to lifting them, and this is especially true when using them on hills. There

The harvest cart at Offbeet Farm is the main way we move veggies from the field to the wash station. *Photo courtesy of Phil Knapp.*

are some innovative designs for well-balanced and ergonomic carts, such as those popularized by Josh Volk of Slow Hand Farm (see resources, page 237). Small trailers pulled by walk-behind tractors, ATVs, lawnmowers, pickup trucks, and full-sized tractors are good options for carrying larger, heavier loads than is possible with hand carts. However, trailers are less nimble than carts, requiring more space to turn around and more skill to run in reverse. Pickup trucks, or any vehicles with beds, are also great options for transporting veggies out of the field. Many farms outfit their tractors or skid steers with forks to either move large pallets or bulk bins throughout the field or to stack the bins on trailers for hauling.

Containers for harvesting and transporting veggies are usually either very small or very large. At the small end, many farms use 5-gallon buckets or bulb crates for harvesting. That's because these containers are relatively cheap and easily moved, stored, and lifted. I personally advocate for using small harvest containers because it places physical limits on how much you can lift at any one time. Even some large storage farms use 5-gallon buckets as direct harvest containers and transfer the veggies into larger containers later. Larger harvest containers are usually bulk, pallet-sized bins such as plastic macrobins. Workers can harvest directly into bulk bins if the bins are driven through the field on forks, or they can empty smaller containers into nearby bulk bins. These are the go-to containers for farms using mechanical harvesters and time-saving equipment like harvest conveyors. Some farmers use the same containers for both harvest and

Harvest conveyors allow crew members harvesting multiple rows to quickly and safely deliver veggies to harvest bins on a trailer without walking back and forth or tossing veggies through the air. *Photo courtesy of Martin's Produce Supplies.*

Harvest Conveyors

Harvest conveyors are useful for more than just storage crops but are very popular among farmers growing lots of cabbage or winter squash. The machines consist of boom arms extending 20 feet or more off a harvest trailer with a conveyor belt that moves crops from the field to the trailer. Usually, several workers harvest along field rows as the conveyor moves slowly down the field. One extra person is required to run the conveyor and direct veggies coming off it into containers. The beauty of this system is that workers don't need to carry (or toss, as many farm crews do) squash or cabbage heads between the rows and the trailer or containers.

storage, while others have specific containers for each task. (See "Storage Containers" on page 47 for more information about different types of containers.)

While your harvest containers will influence your options for handling equipment and vice versa, this relationship isn't equal. I suggest prioritizing handling equipment when considering upgrades. Larger and stronger handling equipment will save you time by reducing the number of trips needed to remove your crops from the field, and, most importantly, large handling equipment can accommodate any container size. This isn't true in the other direction; large containers aren't compatible with small-scale handling tools. Additionally, handling equipment upgrades are often less expensive than container upgrades. This is especially true when purchasing bulk storage bins, which can be several hundred dollars each. Of course, upgrading both containers and handling tools simultaneously is fine, but if you're limited by your budget, purchase better handling equipment first.

Chapter 3

Postharvest Handling

Pulling crops from the field isn't the end of the story! There are additional steps between harvesting, storing, and eventually selling (or eating) our crops to ensure their quality in storage and that our customers (or our families) are happy. These include postharvest tasks like trimming, curing, washing, and grading. Some of these steps are purely for aesthetics or convenience. Other steps, like curing particular crops, will improve both quality and longevity in storage.

Trimming

For the classic root crops, trimming is as simple as removing the leaves. There are several reasons to do this prior to storage. Even though many people enjoy

Trimming root vegetables is much like trimming fingernails. Just as you want to avoid cutting to the quick, you don't want to cut into the vegetable. *Photo courtesy of Phil Knapp.*

eating and cooking with leaves from root crops—for example, beet and turnip greens—these aren't hardy in storage and quickly become sources of rot. While you *can* store crops whole for a short time, the leaves will become unattractive after only a couple weeks. Another reason for trimming off leaves is logistical; leaves add weight and bulk! Why move leaves around and make space for them in storage when it's not necessary?

One last reason to trim off leaves is to prevent them from acting as conduits for moisture loss. When I lived in Sweden, the farmers I worked for had me cutting firewood in the spring. They told me that by felling live silver birch just after leaf-out, when "the leaves are the size of a mouse's ear," and by leaving the trees on the ground for several weeks before cutting and splitting, they can ensure they'll have dry wood by winter. (You'll see this technique mentioned in Lars Mytting's book *Norwegian Wood*.)[1] Trees don't die immediately after felling, just like vegetables don't die right after they're harvested. Those tiny, mouse-ear leaves grow to full size—even on felled trees—as they begin to photosynthesize and transpire water, which under normal circumstances would be replaced by groundwater via the roots. Without that connection, however, the water conduits in the wood are sucked dry, and the wood gets a head start on drying before it's even cut

Rutabaga stems are more elongated than those of most other root crops. Instead of trimming just leaves, you can trim the stems down near the root crown. *Photo courtesy of Phil Knapp.*

and split. Leaves left on root crops after harvest do the same thing, but in this case they pull moisture from the still-attached stems and roots. Trimming off the leaves before storage helps prevent our root veggies from becoming desiccated and floppy.

Ideally, you should leave only the smallest stubs of petioles—the thin stalks that connect the leaves to their stems—behind when trimming. (Even root crops have stems, albeit shrunken ones on the very tops of the root crowns.) The cut petiole surfaces dry out quickly, preventing further moisture loss and stopping pathogens from entering the main root tissue. Cutting too deeply will expose vulnerable inner stem or root tissue. Such cuts tend to leak more moisture and provide direct entry points for pathogens. It doesn't necessarily spell disaster for that veggie, but an errant cut increases the chances that something will go wrong in storage. One exception to this rule is rutabagas. The stems on rutabagas are more elongated than on other root crops—think of the cylindrical and (usually) purple part that extends upward from the root, growing whorls of leaves. It's usually OK to cut into rutabaga stems and even root tissue when trimming. Another exception is allium crops like onions and garlic. With these crops, the portions we eat are actually swollen leaves, so we trim leaves above the swell of the bulbs.

With some crops, trimming can also make washing easier, make the veggies more attractive to customers, and help remove damaged portions prior to storage. I admittedly do this with root brassicas, trimming off portions damaged by cabbage root maggots. I've found that their cut surfaces dry quickly and hold up moderately well in storage. With celeriac, I try to remove the outermost layers of the tangled root mat prior to washing. The same goes for rutabagas, which have lots of small root hairs that tend to hold onto soil and make washing more difficult.

Onions and garlic usually require additional cleaning and trimming after harvest and curing (see "Curing," page 34). This involves removing the outermost, dirty scales, the leftover tops, and the dried roots. Depending on the amounts you grow, this can

Residue Management

You can't talk about trimming without mentioning the resultant crop residues. When I was new to storage farming, I wasn't accustomed to leaving residues in the field and would harvest storage roots tops-and-all, cutting tops as a separate step. I quickly saw the inefficiency of this process. Not only did I haul around more material, I also created large piles of leaves that needed to be dealt with separately. Now, as a rule, I do most of my trimming in the field alongside harvest. This compresses the tasks of harvesting and trimming into a single step: grab, trim, drop; grab, trim, drop; grab, trim, drop. All those nutrients and organic matter stay in the field, where they'll benefit next year's crop. Of the farmers I spoke with, the vast majority operate like me, leaving all their residues in the field and incorporating them the following spring. A handful of farmers had concerns about diseases and pests propagating in the residues and tried to remove all traces of some crops from their fields each fall.

Leaves and stems trimmed from harvested crops stay in the field to become soil organic matter and feed future crops. *Photo courtesy of Phil Knapp.*

be very time consuming. On many farms, onion cleaning is saved as a rainy-day activity during the fall. Though the process isn't quick, it is simple: Slough off the dirty outer scales and trim back the roots and tops. One machine that can speed things up is an onion topper, which consists of spinning cylinders that slowly propel onions along while pinching and removing their tops. (This machine will also remove some fingers if you're not careful.) While toppers will take off some roots and outer scales, most remain behind. There are also dry brush tables that work much like toppers, with brushes in place of cylinders. These effectively remove dirty outer scales and some roots left behind during earlier processing.

Precooling

I don't know many storage farmers who precool their veggies, but it's worth mentioning as a possibility. (Instead, most storage farmers rely on refrigeration systems in their storage rooms to cool crops down to storage temperatures.) Precooling usually refers to rapidly lowering the temperature of vegetables after harvest to extend storage life. With storage vegetables in particular, lowering the temperature quickly slows down the consumption of sugars and production of bad-tasting compounds.[2] For many crops, reducing their temperatures below 40°F (4°C) as quickly as possible after harvest reduces problems with storage rots. Precooling can also reduce the cooling load on your

Taking advantage of a sunny, warm afternoon to dry winter squash at Offbeet Farm before they go inside to finish curing. *Photo courtesy of Phil Knapp.*

refrigeration system and ensure that crops already in storage remain cold. There are special pieces of precooling equipment like hydrocooling tanks, precooling tunnels, and carton coolers. If the weather is warm around harvest time and you're struggling to lower vegetables' internal temperatures quickly enough, these are options to consider. Other strategies include avoiding harvesting on warm days (or at warm times of day), harvesting later in the season when outdoor temperatures are lower, and purchasing a refrigeration system with enough power to handle the cooling load.

Curing

Curing is a general term describing the extra steps that prepare certain vegetables for storage. Most veggies don't need any curing prior to storage. The few that do include winter squash and pumpkins, potatoes, sweet potatoes, and alliums like onions, garlic, and shallots. The conditions needed for curing and the physiological changes that occur are different for each vegetable/group.

For winter squash and pumpkins, curing helps to harden their skin, heal wounds, dry down their stems, and turn starches into sugars like sucrose and glucose. (The correct and lively term for that bit of plant tissue connecting a squash fruit to the vine is *peduncle*.) In some climates, winter squash and pumpkins can cure in the field, but in some regions, farmers need to create the appropriate conditions. Holding squash at 75 to 85°F (24 to 29°C) and 50 to 70 percent relative humidity (RH) for a week or two usually completes the curing process and improves their performance in long-term storage. (Note that not all types of squash need curing, and a separate curing step isn't necessary in all climates. See "Winter Squash and Pumpkins," page 165.)

34

Onions cure in the attic of the storage building at Offbeet Farm, where it's dry and warm, especially on sunny days.

Though potatoes and sweet potatoes are from entirely different plant families, the physiology behind their curing is similar. Both crops go through a process of suberization while curing, whereby a waxy protective layer forms over the skin and wounds. This layer prevents moisture loss and protects against infection from bacteria and fungi. Common wisdom says that skins thicken during the curing process, but in reality, the outermost skin layers just attach more firmly to the tissue underneath.[3] While potatoes can cure under a range of conditions, a temperature between 54 and 59°F (12 and 15°C) is best for speed and preventing pathogen growth. The humidity should be high, at about 80 percent. Sweet potatoes require a similar humidity but much warmer temperatures, ideally between 82 and 86°F (28 and 30°C).

Garlic and onions, including shallots, can cure once they've fully matured and have entered a state of dormancy. Warm and dry conditions over several weeks dry out their outermost scales, which become hard and flaky. Although they can cure in a wide range of temperatures and humidities, optimal conditions range from about 80 to 90°F (27 to 32°C) and 50 to 70 percent RH. The dry outermost layers protect onions and garlic from water loss and pathogens. In both onions and softneck garlic, it's also important that the necks dry and contract during the curing process to prevent the entry of pathogens.

When I was first starting out, curing seemed a bit like alchemy to me. I read the advice from university extension services, but I didn't fully grasp the importance of curing to success in long-term storage. My first failure came with onions, which I grew for my winter CSA during the first year of Root Cellar Farm. I started them from seed and diligently tended them under borrowed grow lights in the living room of my rental house. (My neighbors and landlords definitely thought I was growing something else.) I planted the seedlings and nurtured them all summer long, and they grew into beautifully plump bulbs I was proud of. Had I sold them at a fall farmers market, it would have capped off a successful first season. Instead, I botched their preparations for storage, leaving them too long in the field during a wet fall and inadequately curing them. My CSA customers received less than half of their onions that year because the bulbs grew mold around their necks and rotted from the top downward in storage. In subsequent years, I learned to harvest onions earlier to allow them more time to dry under fans in the garage. I also became cautious of rainfall on onions after their tops fell, a caution I harbor to this day. (My caution may not be entirely warranted, though, as other farmers leave onions out in intermittent rain without issue after tops have fallen—see "West Farm," page 203.)

Curing needs can vary widely from year to year and from place to place. I faced some steep learning curves after moving from Michigan's Upper Peninsula to interior Alaska, and I needed to adapt my curing techniques to the new climate. My biggest challenge with

Winter squash almost never cure outdoors at my farm in Fairbanks, Alaska. Instead, I bring them indoors and cure them at 80°F (27°C) and 50 to 60 percent RH for about a week. *Photo courtesy of Phil Knapp.*

curing in Alaska has been with winter squash. In Michigan, winter squash and pumpkins usually grew to full maturity on the vine—corky stems (*ahem*, peduncles), firm and waxy skins, ground spots, and so on—but in Alaska's abbreviated season, most winter squash barely eke out maturity before frosts arrive. On top of that, the weather around harvest time in late August and early September is typically cool and wet. In Michigan, I had no issues sun-curing my winter squash, either in the field or on the dry and sunny deck. In Alaska, I have to create those conditions indoors.

Wherever you live and whatever your typical weather patterns are, you should be prepared to cure crops indoors if necessary. This can be especially tricky if you grow multiple crops that need curing, and those crops cure under different conditions. In practice, you should identify which spaces are conducive to curing and how different crops might

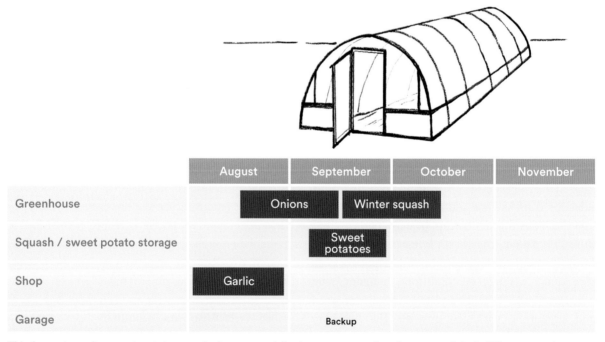

	August	September	October	November
Greenhouse	Onions	Winter squash		
Squash / sweet potato storage		Sweet potatoes		
Shop	Garlic			
Garage		Backup		

This farm stores its sweet potatoes and winter squash in the same room but, because of their different needs, doesn't cure them together. They cure sweet potatoes in the storage room (where it's easier to maintain steady temps and humidity), while winter squash join the sweet potatoes after curing in the greenhouse. They also cure onions in the greenhouse before the winter squash. The farm cures garlic in a shop building and has a garage available as backup in case of a bumper crop or a problem with one of the other spaces.

overlap their time in those spaces. Extra spaces that can be heated and ventilated as backups will help you deal with timing problems or unusually large harvests. It's a helpful coincidence that winter squash and pumpkins and the alliums cure in similar conditions. Many growers use greenhouses to cure these crops since warm and dry environments work for both. Heated shops and sheds can also work for onions, garlic, and squash. Most farmers I know growing sweet potatoes use their storage spaces for the curing period as well, which eliminates extra handling steps between the field and storage. It's common for potato growers to cure potatoes either in ambient conditions (in other words, without manipulating temperature or humidity) or in their potato storage spaces, similarly to sweet potatoes.

You also need to plan out your curing setups and equipment. Airflow is key for all curing crops. The constantly flowing air removes and distributes moisture, keeps temperatures even, and prevents buildup of gases like carbon dioxide and ethylene. Keep this in mind when you choose containers or platforms for curing. I once made the mistake of curing winter squash on plywood shelves. With the impervious plywood and inadequate fans, humidity built up underneath my squash and I had problems with rot on their undersides. Many farms, especially those at small scales, cure alliums and squash in single layers on wire racks or tables to maximize airflow across and under the crops. Some larger farms cure their onions and squash in bulk bins, either singly or in stacks. The trick here is to place fans on top of the bins, lying flat and pointing upward, to pull air up through them; it's almost always easier to pull rather than push air through something. If the fans over the bins aren't powerful enough, you can improve airflow in the stacks by wrapping the outsides, creating makeshift chimneys. Farms that store things like potatoes and onions in bulk piles often use perforated pipes or culverts in the piles for ventilation. Similar to aerating compost, ventilation pipes deliver fresh air to the bottoms of bulk piles to more effectively maintain temperature and control moisture.

How much ventilation is enough? As a rule of thumb, air should be moving, however subtly, at any place in a pile, stack, or bin of curing veggies. Using your hand, you can try to feel air movement under or through stacks or piles of veggies. If you're unsure, you can also use smoke or dust to observe subtle air movement. By observing how the smoke moves—up through the bottom of a stack of bulk bins, for example—you know air is actively moving through your curing crops. If there's not enough airflow, add more or larger fans.

Washing

When I first got into farming for long-term storage, I was surprised to learn that washing veggies upfront, prior to storage, was a viable option. Until then, I'd only ever heard that washing before storage led to poor results. "It's perfectly all right to leave a light coating of dusty soil on the surface of your root vegetables," write Mike and Nancy Bubel in their book, *Root Cellaring*. "Gently brush off excess dirt, but avoid scrubbing or washing the roots. They'll keep better if you don't clean them up too vigorously."[4] This matches most, if not all, advice I've read aimed at gardeners and homesteaders. But then, while learning to grow storage crops commercially, I encountered advice like this excerpt from Ruth Hazzard, Extension Educator for Farmers at UMass Amherst: "One long-standing debate is on the merits of washing carrots before placing them into storage. The decision to wash carrots before storage, or immediately before sale during winter months, is usually based on the farm's washing facilities (heated or not) and available labor."[5] Clearly, there's a disconnect between the two camps, and this discrepancy made me anxious. To wash or not to wash?

I now suspect that both suggestions are correct within their own context. Advice like that offered by the Bubels comes from centuries-old experience and traditions of farmers and homesteaders. Before the widespread use of electricity, fans, mechanical refrigeration, and the like, root cellars, trenches, and cold-holes were some of the only means of preserving fresh vegetables through winter. But these passive means of storage rarely maintain the best conditions for storing vegetables on their own. If temperature and humidity are difficult to regulate, it makes perfect sense to avoid any potential damage and excess moisture from washing veggies before storage. Passive root cellars are often warmer than is optimal, which can allow microbial problems to grow and fester over time. With modern winter storage facilities, farmers can easily regulate temperature, humidity, and airflow. Small bumps, nicks, and excess moisture from washing don't matter as much because lower temperatures limit microbial growth and humidity regulation removes excess moisture. As Ruth Hazzard described, farmers now have the freedom to make decisions about washing based on their washing setups and workloads.

This is a good thing for me and other farmers who can wash veggies most efficiently in the fall! I think about a story Angus Baldwin of West Farm in Jeffersonville, Vermont, told me about trudging his veggies over a half-mile-long snowy path between an old storage shed and washing station (see "West Farm," page 203), and I shudder. I've been there too. During my first winter at Root Cellar Farm in Michigan, I saved all my potato washing for winter and scrubbed them, one load at a time, in a tiny utility sink in my cold basement. That after moving the veggies down narrow stairs and back up if there was any excess after packing. That only lasted one year; after a winter of chilly and slow scrubbing, I washed everything before storage the next fall.

For many farmers, myself included, washing veggies in the fall immediately after harvest is far more feasible than during winter. At Offbeet Farm, the well is up the hill next to the house, about 500 feet from the storage building and the adjacent washing station. In the summer and fall, it's no problem to pipe water down the hill for washing veggies, but problems arise as soon as below-freezing temperatures arrive. There's a small water-holding tank in the storage building for handwashing and miscellaneous rinsing, but serious washing requires more water and space. In the packing side of the storage building, the interior walls and electrical fixtures aren't entirely water-tight, and spraying water with a hose or pressure washer in the small space would be a bad idea. Anyone in a similar situation, where winter water use isn't possible or practical, will need to wash veggies up front. Farms with spacious, heated, and water-resistant facilities with water outlets and proper drains have the option to wash during the winter. They can put crops into storage dirty and clean them up as needed for packing orders.

Besides the washing setup, there are other reasons why farmers might choose either to wash storage crops alongside harvest or wait until winter. The availability of labor around harvest time is a big one. Washing crops might take up precious time you don't have, especially in the fall when there's already so much else to do. That's why some farms put all their fall efforts toward harvesting (see "North Farm," page 177), or wash as much as possible in the fall but leave some washing for the winter (see "Food Farm," page 219). On farms where winter washing isn't possible, washing has to be part of the fall workload.

Some farmers use up-front washing as an opportunity to sanitize certain crops, such as winter squash, prior to storage to lessen the occurrence of storage rots. Janaki Fisher-Merritt at Food Farm in Wrenshall, Minnesota, has recently started sanitizing fall-washed carrots to combat fungal storage rots. There are also aesthetic issues that might influence when farmers wash their crops. With carrots, for example, fine silt or clay soils can easily stain the roots when they're stored dirty for more than a couple months. On the flip side, fall-washed carrots (and other root crops) often develop discolored root tips or fine roots that would otherwise come off during wintertime washing but must be trimmed off if washed in fall.

Which storage crops need washing anyways? Usually, all the classic root crops—beets, carrots, celeriac, parsnips, potatoes, radishes, rutabagas, sweet potatoes, sunchokes, and turnips—are washed to remove soil and expose those beautifully colored skins.

Farmers sometimes wash winter squash and pumpkins to remove soil and sanitize them. As a leafy green sometimes eaten raw, kale should be washed prior to selling for food safety concerns. Cabbages are rarely washed because the outer layers can be peeled off, exposing clean leaves beneath.

Washing setups can be as simple as a slatted surface and a hose. I used the spray-table method on my first, smaller farm in Michigan. The table was a simple frame of 2 × 4s topped with half-inch hardware cloth, and the whole thing sat on sawhorses. I eventually added rails to stop veggies from rolling off the sides. If you've ever used a spray table, you know that washing can be time-consuming and a bit aggravating. You spray and spray until the veggies look pristine, but your heart sinks when you realize they're still dirty on their undersides. It usually takes many repetitions of spray, roll; spray, roll; spray, roll until the batch is entirely clean. Unfortunately, I don't have good notes from those early days regarding my washing speed; I doubt it was more than 100 to 200 pounds per hour on a good day. Of course, there are faster and more efficient ways to wash veggies.

Barrel Washers

Barrel washers are one of the most ubiquitous pieces of equipment you'll see on storage farms. They are exactly as their name describes—barrels that wash crops, usually by spinning slowly while water sprays the tumbling veggies. That annoying problem with spray tables—how the veggies vexingly stay dirty underneath no matter how much you spray them—is solved by barrel washers. They're mainly used for root crops, with the exception of sweet potatoes, which are

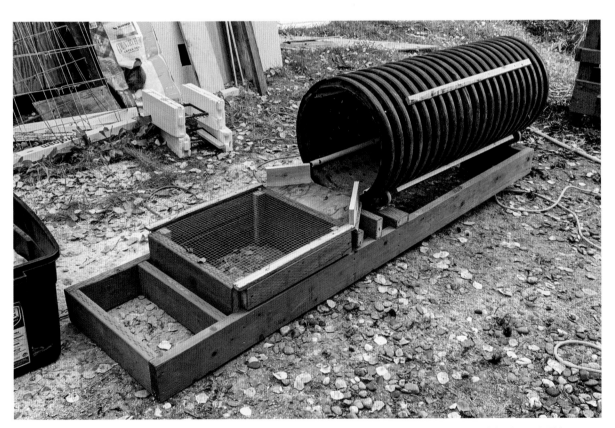

Barrel washer 1.0 at Offbeet Farm was hand-turned using wood purlins fixed to the outside of the barrel. This was a very cheap solution for the washing needs of a budding, cash-strapped farm.

often damaged by the tumbling. Squash are too susceptible to bruising, and I've never heard of someone using a barrel washer for leafy greens.

The beauty of barrel washers is that their design is superbly functional but also adaptable; they're often as unique as the farms on which they're found. They can be small or large, simple or elaborate. The design possibilities are nearly limitless, but most have these features in common: an entry chute of some kind to funnel veggies into the barrel, a barrel with holes or slats for drainage, wheels that fit onto tracks or channels that gird the barrel, a means of turning the barrel, a water source, and something that blocks the exit hole while crops tumble. Here are a few examples that I've made or seen through the years.

In the first year of production at Offbeet Farm, I built a barrel washer using a discarded section of 24-inch plastic corrugated culvert I found in the woods on the property. The design was fairly basic. I built a wooden frame out of treated lumber that included a platform for resting totes, a mini spray table before the barrel, a funnel into the washer, and supports for caster wheels that fit into the barrel's corrugated grooves. I peppered the barrel with ¾-inch holes and fastened pieces of 1-inch PVC, cut in half lengthwise, to the inside of the barrel to act as agitators. I also fastened several wooden purlins on the exterior of the barrel to act as handholds. An old freezer grate on a hinge stopped veggies from rolling out the exit hole during washing. I manually sprayed veggies with a small electric pressure washer. This hand-turned barrel washer was an undeniable improvement over the spray table. It was cheap and easy to build, a perfect combo for a farmer strapped for both time and cash. I used it in this configuration for two seasons before upgrading, sending a little over 10,000 pounds of root veggies through it the second year. There were some obvious downsides to the design after using it awhile. The funnel into the barrel didn't slope downward, and it was time consuming to push veggies into the barrel by hand rather than letting gravity do the work. Hand turning was also tiring on my shoulders!

The second iteration of the barrel washer at Offbeet Farm improved on the design of the first. Modifying the original frame, I removed the spray table and flat funnel to install a sloped chute into the barrel. (This coincided with changing our harvest containers from Rubbermaid totes to 5-gallon buckets, which are lighter to lift.) I used sections of the plastic culvert as hoops girding a circle of wooden slats; the two end-sections of culvert fit the original castor wheels on the frame. A ⅛-horsepower electric DC motor turns a 19-tooth sprocket and ⅝-inch chain wrapping the middle section of culvert (catching the ends of #10 bolts to turn the barrel). A DC variable-speed controller allows us to change how fast the barrel spins. At 100 percent speed, the barrel spins at about 10 RPM, which feels a little fast for most things, and I usually run it between 40 and 70 percent. (The motor has an integrated 40:1 reducing gear, which slows the base rate to 62 RPM.) The new chute makes loading veggies into the barrel much faster, and the motorized turning saves both time and effort. I have chosen, for now, to omit internal water sprayers and use the pressure washer as the water source. The pressure washer effectively cleans the veggies with relatively little water. (Mine uses 1.5 gallons per minute.) A downside is that washing the veggies isn't entirely hands-free.

Giving credit where credit is due, my designs were adapted from plans for a weld-free root washer created from a SARE-funded project led by Grant Schultz of Versaland Farm in Iowa City, Iowa. These plans are available for free on Farm Hack. (If you're not aware, Farm Hack is a place where farmers—or anyone, really—can post and share DIY tool designs in a free, open-source online format. There are many interesting and innovative ideas there. See resources, page 237.) Schultz's barrel washer uses a wider culvert, a more powerful motor, and a larger, 1 ¼-inch pitch chain. I don't think a more powerful motor is necessary—the ⅛-horsepower motor I use delivers plenty of torque—but I'd recommend the larger chain pitch if you can find it. Wider spaces between the links will

The barrel washer built by the carrot royalty of Fairbanks, Alaska, Spinach Creek Farm. Every year, they use this washer to process 30,000 to 40,000 pounds of carrots.

make it easier for the chain to fit over the bolt ends sticking out of the barrel.

Other neat designs I've seen include the barrel washer at Farragut Farm outside Petersburg, Alaska. That washer uses old bicycle rims instead of culvert pieces to gird the wooden slats, and a rubber belt rather than a chain to turn the barrel. The washer is also easy to move around, because the farmers welded the frame onto a set of bike wheels, similar to a garden cart. Spinach Creek Farm, just outside Fairbanks, Alaska, made a barrel washer cleverly powered by a cordless drill, similar to the mechanism on the Tilther tiller sold by Johnny's Selected Seeds. The farmers there created a "floating" block that stops veggies from exiting the spinning barrel, and they made a custom bin flipper to lift and turn their harvest bins at the barrel's entrance.

Of course, there are companies that manufacture and sell barrel washers and other washing equipment. Grindstone Farm makes a wooden-slatted barrel washer similar in design to Schultz's Farm Hack models that's a popular and affordable choice for farmers looking to buy rather than build. Univerco makes both full-sized and mini barrel washers designed for medium to small farms. Other choices include barrel washers from AZS, an Amish-owned company based in Pennsylvania that makes a whole suite of very popular wash-line equipment for small-scale farms. AZS offers full-metal and polycarbonate barrels that integrate into their full set of wash-line equipment. You won't find much of an online sales presence from AZS, but you will find evidence of the popularity of their products. I received a tip for finding AZS products through Nolt's Produce Supplies of Leola, Pennsylvania, which publishes an online catalog with the full line of AZS products (see resources, page 237).

Other Washing Equipment

An increasingly popular piece of washing equipment is the rinse conveyor. These machines move veggies through a series of low- and high-pressure rinses on a slow-moving chain belt. Think of an industrial, through-put dishwasher but for vegetables. Most types have recirculation tanks that allow farmers to use less water and, if applicable, add sanitizers. Internal baffles and screens collect debris and sediment and prevent them from clogging the pumps. Unlike barrel washers, I've never seen a homemade rinse conveyor, but maybe that's because their popularity is relatively new. Rinse conveyors can do everything that barrel washers can do and more, including wash sensitive crops like winter squash and sweet potatoes. I've even heard of farmers using them to rinse very sensitive crops like tomatoes. AZS makes rinse conveyors in several widths, depending on your needs, and they're very popular among storage farmers, including several profiled in part 4 of this book.

Brush washers are also popular among some storage farmers. Designed mainly for removing dirt and stains from durable root crops, brush washers send veggies over a series of quickly spinning brushes. I've heard of several farmers who use them to remove soil staining from carrots that were stored dirty for too long. Brush washers get a bad rap from the food safety community because it's difficult to fully clean and sanitize the brushes between washings. Vegetable parts get easily stuck between all those tiny bristles.

Dunk (or soak) tanks can be used in conjunction with any other washing equipment discussed here. The idea is simple: presoak your veggies to loosen and remove some soil before sending them into your primary washing equipment. Some soak tanks are integrated parts of a wash line, with conveyors that remove veggies from the tank and onto the washer. Dunk tanks can also be as simple as totes or livestock waterers large enough to accommodate harvest totes. At Offbeet Farm, we've used Rubbermaid totes to presoak our 5-gallon buckets (peppered with drainage holes) prior to washing.

Grading

Grading is usually the last postharvest step before you drop veggies into storage containers or pack them for sale. Separating crops into grades is a useful marketing

The Food Farm crew grades potatoes coming through the brush washer and onto a roller table, both of which are made by AZS. *Photo courtesy of Janaki Fisher-Merritt.*

strategy, but importantly, it can also help a farm maximize its available storage space. The act of grading is easy: Separate the pieces into different categories (or grades). Making decisions about the different grades and following through on separating, labeling, storing, and selling them takes more forethought and planning.

Grade criteria for each vegetable are unique to individual farms. At the very basic level, there are two grades: salable pieces and culls. Consider your customers' standards, your own standards, and food-safety requirements. For example, will you trim off the rodent bites from a carrot crown or cull the whole carrot? Does one wireworm hole ruin the whole piece, or will your customers be accepting of this? It's a good idea to ask your customers and other farmers for their opinions, but you have to make the final decisions. You also need to decide what to do with culls. Many farmers choose not to return anything with disease issues, including storage rots, to the fields. They either feed these to local livestock, such as pigs, or heap them into compost piles used for landscaping or other projects.

I also set aside seconds, an intermediate grade between culls and my regular sales. These aren't rotted but are either highly trimmed or too weird to sell normally. Besides eating them myself and giving them to crew members, I sell seconds at discounted prices at the end of the storage season and donate them to my local soup kitchen.

Your regular salable grades will depend on your sales outlets. For example, if you sell to restaurants, you may be meeting the specifications of individual chefs. Steve Pincus of Tipi Produce in Evansville, Wisconsin, told me a story about grading carrots to sell to a local juicing bar, who said they only wanted the largest and straightest carrots because those were easiest to shove into a juicing machine. While selling wholesale to local groceries and co-ops, I've heard repeatedly from produce managers that loose, bulk veggies should be the highest-quality, most aesthetically pleasing pieces. Customers are more forgiving of weirdly shaped vegetables when they're pre-bagged, but they are more discerning when selecting individual pieces for themselves. Some customers also prefer veggies of similar size, whether in bulk or bagged. Small pieces are inevitable, and instead of culling these in the field or including them with larger veggies, I set aside small pieces to sell together as "minis" or "babies." It always surprises me that small beets, for

Paddling beets from the washer, to the grading table, and to a waiting bin at Offbeet Farm. *Photo courtesy of Phil Knapp.*

example, might annoy customers when mixed into a bag of larger beets but delight them when sold together as "baby beets."

When and where should you grade your crops? Some farmers, myself included, try to do as much field culling as possible. This stops me from transporting and processing pieces that won't ever make the cut, unnecessarily taking up space and removing nutrients from the field. In the case of certain insect or disease damage, you may want to remove those pieces from the field anyway to reduce the chances of further spread. Depending on your harvest methods and machinery, you may not have a chance to inspect a crop alongside harvest. Beets or carrots flying out the back of a Scott Viner harvester, for example, sail by too fast to grade. Even for farmers working at hand scale, it's not possible to field cull everything since problems are often hidden by soil and debris.

After harvest and field culling, most grading happens either while washing or before packing for sale (depending on if and when a farmer washes a crop). I grade out culls during fall washing, but I wait until packing orders to separate my salable grades. This is how most farmers who wash veggies in the fall do it. Farmers who wash in the wintertime usually do their salable grading while washing crops for sale.

I use a grading table as part of my wash line. Grading tables are places to spread out and inspect washed veggies before packing them into containers for storage or sale. My grading table is simple: It's an old section of composite countertop with wooden rails fixed to two sides. Any surface that's water resistant and easy to clean and sanitize will work, but one that's smooth and slippery, for veggies to slide over quickly, works best. You can also purchase grading tables that integrate into wash lines. For example, AZS sells a popular model that's circular and slowly spins, bringing veggies from the washer to waiting workers.

You can also grade directly from storage containers before sale. I pack one grade for sale while keeping extra containers for my other grades, including culls, nearby. Whenever I see a piece that fits those other grading categories, I put it in the appropriate container. Those boxes or bags with other grades usually go back into storage labeled, to be packed up another day. One other grading method is to leave certain grades behind in a storage container by selectively removing other grades. You might, for example, remove your salable parsnips while leaving seconds and culls behind. Eventually, only weird pieces will remain in the storage container. This is, in essence, what Jonathan's Farm near Winnipeg, Manitoba, does when they bring their storage bins to free-choice CSA pickups. The CSA customers select out the pieces they want, and eventually, only the dregs remain (see "Jonathan's Farm," page 191).

Chapter 4

Into Storage

After all the harvesting, trimming, curing, washing, and grading, it's time to put those veggies into storage. Just toss them in, right? Not exactly. As with other aspects of storage farming, there's some logistical planning that will ease the process of actually placing crops into their winter abodes. Part of that is ensuring you're set up beforehand—that you're prepared to give the various crops what they need in terms of storage conditions and space. The first year I stored crops commercially, I did not give enough thought to storage containers nor how to arrange them in my cold room. I forced myself to make decisions on the fly, some of which I regretted later in the winter.

There's also that critical period just after crops go into storage. In the first days and weeks, the veggies will show signs of softening if the air is too dry and might begin sprouting if the room is too warm. Condensation will form inside containers and on the crops if there's inadequate ventilation. It's crucial to be observant during this time and to make changes if necessary. By acting quickly, you can usually remedy small problems before they balloon into insurmountable ones.

Storage Room Conditions

Success in long-term storage comes down to nailing the storage conditions, of which temperature and humidity are the most important. The goal is to maintain dormancy in our crops while minimizing decay and desiccation (moisture loss). If you grow many different crops, then it might seem like a daunting task to provide the right temperatures and humidities for all of them.

Lucky for us, many common storage crops prefer similar conditions. Depending on what you grow and store, though, you might need to provide four or five different sets of conditions to best meet your crops' needs. If that's not possible, you'll need to decide which crops get ideal storage conditions and which crops don't.

The Cold and Damp Group
32°F (0°C) and above 98 percent RH
Many classic storage crops keep best in cold and damp conditions. By cold, I mean as close to the freezing point as possible, right at 32°F. This might seem precarious, but basic chemistry provides a safeguard. The freezing point of water decreases when you dissolve things in it—things such as salts and carbohydrates in vegetables. The average freezing point of carrots and beets, for example, is actually around 29°F (−1.7°C).[1] Keeping the storage temperature at 32°F gives you a small buffer against inadequate airflow and cold spots, as well as power outages should they happen. It also prevents any water used for humidifying the space from freezing. As for relative humidity, damp means that it should be at or above 98 percent.

Most of the root crops fall into this group, including beets, carrots, celeriac, parsnips, radishes, rutabagas, sunchokes, and turnips. So do leafy and fleshy brassicas, including Brussels sprouts, cabbage, kale, and kohlrabi. Some homesteaders advise storing brassicas and other crops separately to avoid imparting a cabbage-like flavor to other veggies. I've never had an issue with this despite many years of storing these crops together. In fact, I don't know of any commercial farmers who separate these crops for the purpose of odor protection. Perhaps it becomes an issue at relatively warm storage temperatures.

The Alliums
32°F (0°C) and between 60 and 70 percent RH

Classic allium crops, including garlic, onions, and shallots, have similar responses to different temperatures and humidities in storage. Alliums can handle a range of storage temperatures above 32°F and still maintain acceptable storage lives. What's crucial is keeping them dry, preferably between about 60 and 70 percent RH. Despite what you might read or hear, alliums store best when kept cold, around 32°F. There's a lot of research to back this up, as well as firsthand accounts from farmers. Temperatures near freezing inhibit both shoot and root growth—which are the main reasons for moisture loss that leads to squishy bulbs—while also slowing down pathogens.

A problem shared by many storage farmers who grow a diverse set of crops is creating a separate, cold-but-dry storage space for the alliums. Many of us don't grow alliums in large enough quantities to justify this, and we store our alliums at less-than-optimal temperatures in relatively dry hallways, sheds, barns, or pack-room corners without modifying temperature too much. So, what happens at higher temperatures? Very warm temps (above 86°F, or 30°C) also inhibit both sprouting and root growth. At these high temps, the relative humidity needs to be around 80 to 90 percent to prevent drying, but this combination can also promote pathogen growth.[2] Most farmers, though, aren't regularly heating spaces into the 80s and above.

Garlic and onions respond differently to storage in intermediate temperatures. If well cured, onions usually do well in temperatures up to about 50°F (10°C), but you'll see earlier sprouting and rotting the warmer it gets.[3] Garlic, on the other hand, sprouts quickly when stored between 40 and 50°F (4 and 10°C)—around three to four months after harvest, depending on the variety.[4]

Potatoes
38 to 45°F (3 to 7°C) and above 95 percent RH

Potatoes store best in damp conditions above 95 percent RH, but the optimal storage temperature depends on your goals. Potatoes are in their own storage group because cold-storage temperatures—near freezing, from 32 to 35°F (0 to 2°C)—can injure potatoes over time, manifesting as browning flesh and surface mold, and temperatures above 45°F (7°C) lead to early sprouting. Since most small-to-medium farms aren't selling potatoes for fries or chips, their ideal storage temps range from 38 to 45°F.[5] (Sources differ on the lower end of the optimum range; I've gone for the middle ground with 38°F.) At the lower end, potatoes will acquire more sweetness and may excessively brown when fried, but lower temperatures tend to increase storage life. At the higher end, the potatoes will convert less starch to sugar (preventing browning when fried), but they will also sprout sooner.

Sweet Potatoes
57 to 60°F (14 to 16°C) and between 85 and 90 percent RH

Sweet potatoes, unlike the other root crops, cannot tolerate low temperatures. They start accumulating chilling injury below 54°F (12°C). It only takes two to three weeks at 45°F (7°C) for outward signs of chilling injury to show.[6] The best storage temperatures appear to be between 57 and 60°F, and humidity between 85 and 90 percent toes the line between preserving moisture and preventing mold growth.[7] Storage temperatures above 66°F (19°C) will lead to excessive sprouting after a few months.

Winter Squash and Pumpkins
50 to 55°F (10 to 13°C) and between 50 and 70 percent RH

Winter squash and pumpkins, despite their ubiquity, are tropical plants and cannot tolerate low temperatures in storage or otherwise. Their sweet spot is between 50 and 55°F. At lower temperatures they accumulate chilling injury, which usually shows first as pockmarked skin. Higher temperatures speed up their deterioration and cause green varieties to yellow.[8] The lower the storage temperature, the faster chilling injuries develop; at 41°F (5°C), chilling injury begins showing after about a month, whereas some squash stored at 50°F (the low end of the optimum range) might show signs of injury after several months.[9] Humidity should be between 50 and 70 percent, but there are trade-offs at each end of the range. Relative humidity closer to 50 percent leads to more moisture loss, while RH closer to 70 percent leads to higher losses from storage rots.[10]

Storage Containers

Storage containers come in many shapes, sizes, and materials, but there are a few traits that good storage containers have in common. First, they should allow some air in and out. Using airtight containers is a bad idea. Many storage farmers, myself included, have tried it, and the results are almost always slimy. The amount of air exchange needed depends on the crop and the humidity in your storage space, and using plastic liners can limit air exchange in containers with too much ventilation. Good storage containers are easy to load and unload, and they're usually stackable to take advantage of vertical space. They're also easy to clean when veggies inevitably go bad inside them.

Harkening back to the section on materials handling (see "Materials Handling at Harvest," page 27), your storage containers should be compatible with your means of moving them. As a result, most storage containers are either small enough to lift by hand (when full) or are very large, requiring handling equipment like forklifts or pallet jacks to move around. Most storage containers are also designed with pallets in mind. Because pallets and pallet equipment are so ubiquitous, having storage containers compatible with pallets—or with integrated pallets—makes finding and using handling equipment easier.

Bulk bins are by far the most commonly used containers on storage farms. Many farmers refer to plastic bulk containers as macrobins. MacroBin is actually the trade name for plastic agricultural bins made by the IPL

Macro company, but I, like many farmers, use the term to refer to any plastic bulk container. Macrobins come in several sizes, mostly variable by height since their width and length usually match standard pallets. Most macrobins have ventilation holes cut into the bottom and sides, and some have side access ports. Their bottoms are generally compatible with pallet-lifting equipment. Wooden bulk bins are also common, and many farmers choose to build their own to save on costs and customize design. However, wood is a more difficult material to clean and sanitize, and it deteriorates through the years. That said, wooden bins are still quite durable and will last for decades if built well (see "Tipi Produce," page 211). One solution to the challenge of cleaning wooden bins is to use plastic bin liners inside of them.

Large bulk bins also come in less rigid, less durable, and less expensive materials. One option is the Gaylord box, named for the Gaylord Container Corporation, which designed and made boxes of this

Not all farms choose to purchase macrobins. Tipi Produce made these durable wooden bins that have lasted nearly two decades. *Photo courtesy of Beth Kazmar.*

Stacks of macrobins in the Boldly Grown Farm warehouse. These bins are stackable, durable, easy to clean, and are probably the closest thing to an industry standard. *Photo courtesy of Amy Frye.*

style in the early 1900s. Gaylord boxes are large card-board containers set on pallets, and you'll commonly see these boxes holding things like watermelons and pumpkins at supermarkets. Depending on how many layers of cardboard comprise their walls, they can hold loads of over 1,000 pounds (454 kg) and stack upon each other. Some Gaylord boxes are even made of corrugated plastic for more durability and easier cleaning. As you can imagine, cardboard boxes are easily dirtied and don't respond well to high humidity. But they're relatively cheap compared to other bulk containers.

Bulk woven bags, generally called super sacks, are another large but inexpensive option. Super sacks are commonly used for holding and storing grain, but they can be used for vegetables as well. It's a good idea to set them on pallets, but they have strong fabric loops on top so they can be lifted with pallet forks. They're commonly rated to hold over 2,000 pounds (907 kg) each. A big downside of super sacks is that they can't be stacked without using pallet racking or other supports. They also lack ventilation holes on their sides and bottoms. However, at less than 10 percent the cost of macrobins, they're the cheapest option for large containers. A workaround for the stacking problem is to use super sacks in conjunction with IBC totes. IBC—or intermediate bulk container—totes are mostly used for shipping and storing liquids and powders. They consist of an open-topped metal cage and pallet with a large (usually around 250-gallon, or about 946-liter) plastic container inside. I don't recommend using the plastic portions since they're often used for shipping chemicals and fuel, but when empty, the metal cages are perfectly sized for super sacks and allow you to stack them. IBC totes are easy to find used, and combined with a super sack, they might be half the cost of a comparable macrobin.

It's worth mentioning again that these pallet-sized options often require large handling equipment: pallet jacks at the very least and preferably stackers or forklifts. However, their sheer size makes these large containers more efficient to handle than smaller containers. Whereas 1,000 pounds of beets might take twenty

In the absence of large handling equipment at Offbeet Farm, we have to fill and empty our bulk super sacks in place, one small container at a time. While it drives me crazy at times, this has enabled the farm to reap the benefits of large containers without the expense of pallet lifters. *Photo courtesy of Phil Knapp.*

smaller hand-sized containers—that must be loaded and moved one by one—this amount can easily fit inside one or two bulk containers.

There is a workaround to the necessity of handling equipment, albeit an inefficient one. It's always possible to load large bulk containers that are already in their final resting place within a storage room. That means filling them from smaller containers dumped one by one, but it gives farmers the space and cost efficiency of large containers without large handling equipment.

Smaller containers that can be lifted by hand are another option for storing winter vegetables. Some farmers use them to store the entire harvest, while others, myself included, use them to store specialty crops in smaller amounts. Plastic totes are common on many farms, especially the 18- and 27-gallon sizes (about 70 or 100 L). (Standard Rubbermaid totes are 18 gallons.) Don't be fooled into thinking that 27-gallon totes are small based on their appearance; they can easily weigh over 100 pounds (45 kg) when full. To make them suitable for winter storage, you'll need to pepper them with holes so they're breathable. If you don't, condensation will build up on the inside of the

Woven poly bags similar to 50-pound grain bags work well for holding in moisture while allowing some air exchange. They're also stackable when arranged carefully in a crisscross pattern. *Photo courtesy of Jonathan Stevens.*

containers and the veggies will become oxygen-starved. The number and size of ventilation holes depends on the crop and the humidity in your storage space.

Similar to totes, bulb crates are another small option. They are durable, stackable, and versatile containers, usually compatible with pallets when stacked into the right configuration. A downside with bulb crates is their breathability in storage spaces that are too dry. As a rule, only use open bulb crates if the humidity in your storage room is high enough. Small woven bags are also common inexpensive storage containers. Burlap and loosely woven poly bags are useful for storing crops, such as onions and garlic, that appreciate dry spaces. Farmers often use these bags for

potatoes and sweet potatoes, too, when the humidity is sufficiently high. It's more common to put root crops into tightly woven poly bags, similar to those used for seed, because they're better at holding in humidity while allowing a small amount of air exchange. Small bags are great because they rarely exceed 50 pounds (23 kg), and they're easily stacked on pallets.

Some farmers skip containers entirely and opt for storing veggies in bulk piles. This is common practice on very large farms growing huge amounts of a single crop. Very large bulk piles need special ventilation, usually through culvert-sized perforated pipes that deliver fresh air to the bottoms of piles. Huge piles like these also need special conveyors to get crops in and out. I've never seen diversified vegetable farmers use bulk piles in this manner. I have, however, seen many examples of bulk piles in corral-style enclosures, typically three-sided and made of slatted wood. As a farmer builds the pile, they add a fourth slatted wall incrementally as the pile grows. Slatted floors and space around the outside of these enclosures can encourage air movement through and around the piles. Bulk piles in corrals are a less expensive option for storing large amounts of veggies, but loading and unloading them can be difficult and inefficient.

Storage Container Needs

Imagine that you're cruising through the harvest, happily washing veggies before putting them into storage, but you notice that your pile of storage containers is shrinking at an alarming rate. At some point, it becomes abundantly clear that you're going to run out. Now what? Looks like you're making an emergency trip to town or calling in that favor with a farmer friend. Wouldn't it be nice to avoid this problem entirely, though?

At some point in the planning process, you should consider the type and number of containers or piles you'll need in your storage space. With certain containers in mind, the goal is to estimate the number you need to store your planned harvests. At the outset, you will need to know the volume of the containers—including corrals for bulk piles—the bulk densities of the crops you're storing, and your planned harvest amounts (see "Setting Goals," page 65).

Table 4.1 contains the bulk densities of common storage crops compiled from members of the University of Vermont Extension and supplemented with data from my own farm. You can use these numbers to estimate the storage capacity (in pounds) of your storage containers. To do so, multiply a crop's bulk density (in pounds per cubic foot) by your container's volume (in cubic feet). Keep in mind that these bulk densities are just estimates; they can vary based on factors such as the size of veggies, how they're packed into containers, and the size of those containers. To make your estimates

conservative—and thereby overestimate the number of containers you need—err on the low side for bulk factors such as the size of veggies, how they're packed into containers, and the size of those containers. To

Density vs. Bulk Density

Density and *bulk density* are not the same thing. *Density* depends strictly on an object's mass and intrinsic volume. *Bulk density* depends on the mass and volume of a collection of objects, including the empty spaces between them. For example, density would depend on the weight and volume of a single apple, whereas bulk density would indicate the weight and volume of an entire basket of apples.

Table 4.1. Storage-Crop Bulk Densities

Vegetable	Pounds per ft³
Beets	27–36
Carrots	26–32
Cabbage	19–25
Celeriac	16
Garlic	24
Onions	22–23
Parsnips	20–24
Potatoes	24–37
Rutabagas	27–34
Squash (Butternut)	31
Sweet potatoes	22–27
Turnips	26–33

Source: Andy Chamberlin, "Bins, Buckets, Baskets & Totes," UVM Extension Ag Engineering Blog, November 14, 2018, https://blog.uvm.edu/cwcallah/2018/11/14/bins-buckets -baskets-totes.

conservative—and thereby overestimate the number of containers you need—err on the low side for bulk density. Also keep in mind that it's not always practical to use the full volume of some containers. Containers with lids—Rubbermaid totes, for example—may lose 5 to 10 percent of their usable volume with the lid on. Lastly, divide the planned harvest weight for each crop by the container's capacity to determine the number of containers you'll need.

As an example, imagine you're planning to use the 27-gallon tote bins commonly sold at hardware stores as containers to store 5,000 pounds of carrots. Each tote has a total volume of 3.61 cubic feet but loses about 10 percent with the lid on. As such, each tote has a usable volume of about 3.25 cubic feet (3.61 ft^3 × 0.9). The approximate bulk density of carrots is 29 pounds per cubic foot, so each container will hold about 95 pounds (3.25 ft^3 × 29 lbs per ft^3). To store your expected harvest of 5,000 pounds, you'll need at least 53 totes (5,000 lbs ÷ 95 lbs per tote).

Remember, actual yields don't always match what you've planned for. Some years yield less than expected, whereas other years yield far more. It's a good idea to have enough containers on hand to accommodate bumper crops. Even if some remain empty, you'll appreciate having more than you need instead of scrambling to find more. How many extras should you have? That's hard to say, but having 10 to 20 percent more capacity than needed isn't a bad idea. In the above example, that would mean an extra five to ten totes.

Arranging Storage Containers

One of the trickiest parts of storage farming is arranging the storage rooms such that all the veggies are accessible when you need them. This is particularly problematic when your containers are stacked into multiple rows. If you're not careful while putting veggies into storage, you'll be presented with a frustrating game of Tetris every time you need to access something new. In the best-case scenario, all the crops you grow will be easily accessible within the storage room. This is a simple matter if you grow just a few crops and varieties or if

A 27-gallon tote can store between 95 and 110 pounds of beets when full (depending on the size of the beets).

you have extra space. However, this takes some careful planning if you're a highly diversified farm packing your storage rooms to the gills each fall. It might not even be possible depending on the number of crops you grow.

Before harvest time, I sit down with a diagram of all the available spaces in my storage room. I first estimate how many containers I'll need for each crop and/or variety that I'm storing separately. To do this, you'll use your own yield projections and estimates for the holding capacities of your containers (discussed above). Table 4.2 shows some example calculations for a farm using typically sized macrobins (32.4 cubic feet). Then, I go through the tedious task of assigning spaces to each crop and variety. I say tedious, but it's actually kind of fun, like a farm sudoku! I highly recommend doing this with something erasable; you won't get it right on the first try. As you're playing this game, it's important to follow practical rules for loading and unloading the space. Pretend that you're actually placing and stacking containers. You can't, for example, magically move a new container to the bottom of an

A full cold-storage room at Offbeet Farm. The access aisle is just wide enough for a rolling cart, making it easy to move small bins of veggies to the pack room when prepping orders.

Table 4.2. Bin Requirements Example

Crops in Order of Harvest	Expected Yield (lbs)	Approximate Capacity of 32.4-ft³ Macrobins (lbs)*	Number of Macrobins
Beets	9,000	925	10
Kohlrabi	7,200	925	8
Turnips	6,000	925	7
Cabbages	13,500	750	18
Potatoes	7,500	950	8
Carrots	12,750	850	15
Parsnips	4,000	675	6

* These capacity estimates assume we're only using 90 percent of the bin volume (to avoid damaging veggies while stacking) and that bulk densities are in the range of those presented in table 4.1 (page 52).

existing stack or put something in a blocked back row without making rearrangements. (See "Storage Room Minimum Dimensions," page 102, for more on height considerations for stacks.)

I also like to include numbers to indicate the harvest and loading order. Definitely take your planned harvest order into consideration and adjust if necessary (see "Harvest Order," page 16). For example, I changed my harvest order last season, harvesting turnips before rutabagas and beets so they could have an easily accessible spot in storage. Once complete, you should have easy access to all crops and varieties you need at the beginning of the sales season. Make sure you maintain the same level of access throughout the winter, as containers empty. You might need to add a layer of

complexity and put multiple crops or varieties inside a single container, in which case you should follow the same rules for loading and unloading. Be sure to take notes while you're actually loading the storage room to update your storage-room map. This will save you from scratching your head months later, wondering where you stashed that one particular variety.

4–Beets	8–Beets		72–Parsnips
3–Beets	7–Beets		71–Parsnips
2–Beets	6–Beets		10–Beets
1–Beets	5–Beets		9–Beets
14–Kohlrabi	18–Kohlrabi		22–Turnips
13–Kohlrabi	17–Kohlrabi		21–Turnips
12–Kohlrabi	16–Kohlrabi		20–Turnips
11–Kohlrabi	15–Kohlrabi		19–Turnips
26–Cabbage	30–Cabbage		34–Cabbage
25–Turnips	29–Cabbage		33–Cabbage
24–Turnips	28–Cabbage		32–Cabbage
23–Turnips	27–Cabbage		31–Cabbage
38–Cabbage	42–Cabbage		54–Carrots
37–Cabbage	41–Cabbage		53–Carrots
36–Cabbage	40–Cabbage		52–Carrots
35–Cabbage	39–Cabbage		43–Cabbage
47–Potatoes	51–Potatoes		58–Carrots
46–Potatoes	50–Potatoes		57–Carrots
45–Potatoes	49–Potatoes		56–Carrots
44–Potatoes	48–Potatoes		55–Carrots
62–Carrots	66–Carrots		70–Parsnips
61–Carrots	65–Carrots		69–Parsnips
60–Carrots	64–Carrots		68–Parsnips
59–Carrots	63–Carrots		67–Parsnips

A sample diagram for planning macrobin arrangement in a cold-storage room using the harvest order and bin estimates from table 4.2. The rectangles represent the footprints for macrobins in stacks of four, and the numbers indicate the loading order. The goal is for all crops to be immediately accessible following harvest and to remain accessible throughout the winter.

Loading Storage Containers

Once you have a plan, it's time to load veggies into their storage containers, which can be a bit frightening. It's like sending them away on a long trip, since you might not see them again for months. If there's one reason to be diligent in the field and at the grading table, it is to prevent bad actors from entering the scene and corrupting your wholesome crops. Apart from removing the obvious sources of rot, there are a few things you can do to maximize your precious veggies' chances of success in storage.

First, be as gentle as possible when handling and loading the veggies. Minimize drop heights, look for and eliminate sharp edges and rough surfaces, and handle the veggies with care. For example, wearing gloves when you reach into a pile of crops will prevent your fingernails from damaging them; try grabbing some beets from a pile bare-handed and look at your nails to see what I mean. At Offbeet Farm, I've used a spare piece of foam board as a ramp to slide veggies into containers from the grading table to eliminate a 10- to 12-inch drop. It's common for farmers using mechanical harvesters to lower the conveyor arm near the bottom of a bulk bin at first, raising it slowly as the veggies fill the container. This reduces the distance the veggies have to drop. Another strategy is to provide something soft for veggies to drop into, such as peat moss or burlap. Porous materials like these also provide some moisture management (see "Ventilation vs. Desiccation," page 56).

Some veggies are more sensitive to nicks and bruises than others. Rutabagas are a good example of a resilient veggie: Don't go carelessly tossing them around, but they can take more abuse than beets, for example, without bruising. Farmers growing sweet potatoes report that they're extremely sensitive to bruising. With long veggies like carrots and daikon radishes, breakage is the primary concern, though different varieties are more or less sensitive. Parsnips tend to bend rather than break despite their length.

If you're washing veggies before long-term storage, it's good practice to allow them to dry fully before

Almost anything you have on hand can serve as a ramp to slow veggies as they fall into containers. Here, Food Farm uses a boogie board for carrots coming off the wash line. *Photo courtesy of Janaki Fisher-Merritt.*

partially enclosing them for moisture retention. An effective way to do this while still cooling them quickly is to place washed veggies into cold storage fully exposed, regardless of the humidity, for a day or two. This means leaving off container lids, plastic liners, or whatever else you're using to limit moisture loss until the veggies' surfaces are fully dry. In long-term storage, liquid water is not your friend.

Lastly, you want to be sure that veggies can cool quickly once inside the cooler, especially when using bulk containers. Excepting alliums, potatoes, sweet potatoes, and winter squash/pumpkins, you should cool veggies below 40°F (4.4°C) within 24 hours of harvest. There are a couple of things that can help remove field heat faster. First, be sure there is space for airflow between storage containers to remove heat via convection. For pallet-sized containers, leave at least 4 inches (10 cm) of space around the sides of each container.[11] Use well-vented containers and avoid bin liners until the veggies have cooled. Cooling veggies from 60°F to 40°F inside a ventilated macrobin—in a 35°F cooler with a correctly sized refrigeration system—will take approximately 24 hours, but the same process might take several days inside a bulk container that impedes airflow.[12] If you're still struggling to cool veggies quickly, it's probably a sign that your refrigeration system is underpowered (see "Cooling Loads," page 109) or needs maintenance or repair. In the meantime, try delaying harvest until colder weather arrives or, if that's not an option, cooling crops inside smaller ventilated containers for a few days before transferring them to larger containers. Precooling these veggies with water or other equipment is also an option (see "Precooling," page 33).

Ventilation vs. Desiccation

Something storage farmers quickly realize is that there's a balance between providing vegetables with enough fresh air (to bring in oxygen and release carbon dioxide, ethylene, and water vapor) and drying them out. Lots of farmers struggle to keep humidity high enough for veggies to sit out in the open without becoming floppy, but sealing veggies without enough air exchange leads to condensation, mold, and worse. So, what's the right balance and how do you find it?

The "easiest" answer is to ensure adequate humidity in the storage room (see "Humidity Control," page 119). If increasing humidity in the entire storage room isn't currently an option, then there's no easy answer, and it comes down to experience and feel.

For farmers using bulk pallet bins, there are two common strategies for limiting moisture loss with plastic perforated bin liners. One method involves using liners that are too tall for your bins. With the perforated liners inside the bins, you can fold over the excess liner to loosely cover the open tops. The idea here is to provide enough airflow to prevent condensation without providing so much airflow that veggies near the opening dry out. If your bins aren't in stacks, you can monitor them and open or close the liners as

needed. If the bins are in stacks, you'll have to hope for the best. The other method is to use a plastic liner (perforated or not) upside down over the outsides of the bins. This leaves the (typically) ventilated bin bottoms open to air exchange while preserving moisture throughout most of the bins' interiors. With super sacks, the woven plastic is usually impervious to much air or water movement, but you can make adjustments by opening and closing the flaps on top of the bags.

It's less common to use plastic liners with smaller containers, so the trick is to make sure the containers themselves provide the right amount of ventilation. For bulb crates and nested harvest crates, this isn't really possible. They are what they are. You can use small food-grade plastic bags and fit those inside the containers, though, and you can place large plastic bags upside down over stacks of containers. Totes such as Rubbermaid bins need to be poked with holes for ventilation. In my experience, larger holes are better than smaller holes. You can use the lids to make micro-adjustments when the bins are full of veggies. (Don't try drilling holes with veggies in the containers; you will stab them with the drill bit.) Farmers using small woven plastic bags usually seal them with good results, seeing little in the way of desiccation or molds from condensation (see "Jonathan's Farm," page 191).

Some of you might be wondering why we're not talking about storing veggies in damp sand, sawdust, or the like. These materials are tried and true for keeping veggies at the right humidity in storage, but the issue is efficiency. It's fine to load and store a few hundred pounds of veggies this way, but it stops making sense at commercial scales. If anyone has an efficient way to load, unload, and clean 10,000 pounds of carrots from damp sand, I'm all ears.

Instead, it's common practice to add a small amount of absorbent material to the bottom of storage containers to help regulate moisture. If you wash veggies prior to storage, this material will absorb some of the excess moisture from washing. If you don't wash veggies before storage, be sure to dampen the absorbent material first, or it will pull moisture from the veggies near it. I've tried all sorts of materials through the years: brown paper, sawdust, wood shavings, burlap, and even paper towels. Unfortunately, burlap, peat moss, and certain types of sawdust and shavings can impart off-flavors to the veggies they touch. Using something like row cover or paper to separate veggies from a more absorbent but messy material will save you the headache of cleaning off veggies later and provide a barrier against unwanted flavors. I like using old row cover as a barrier because it's odorless and it gives me an extra use for ragged pieces before I throw them away. Floating row cover is made from woven strands of polypropylene—the same plastic used in yogurt containers—which the FDA considers to be food-safe and which is inherently BPA-free. I select only clean pieces of row cover and wash them thoroughly first.

Bulk bins covered with upside-down liners at North Farm. Some veggies toward the bottom might dry out, but the liners will prevent entire bins from "liquefying," as is common when they are completely sealed. *Photo courtesy of Allison Stawara.*

On Offbeet Farm, we line all our storage containers with a layer of peat moss covered by clean pieces of floating row cover.

Be diligent about moisture management in storage. Getting it right takes observation and the ability to change tack if needed. Whenever I'm in my cold-storage room, I'm touching the veggies to check for softening and looking for signs of condensation and excess moisture. If there's a problem, I may adjust the humidity level in the room, adjust the ventilation on certain containers, or even temporarily throw a plastic sheet over the troubled veggies. Once they're severely desiccated, there's no going back for most vegetables, and those pieces end up in my seconds container. Even though I'm on a personal mission to convince the world that slightly floppy veggies taste incredible, we're stuck carefully managing moisture in our winter storage rooms until people see the light.

Part 2

Designing a Storage Farm

Farming for winter storage frees up time in the summer for tending crops that would otherwise go toward prepping and making sales.

Chapter 5

The Business of Storage Farming

As I described in the introduction, I was surprised when first encountering the enthusiastic demand for locally grown veggies during the winter. In many areas, winter storage can address a gap in the local food system. If you walk into any grocery store in the northern United States and Canada in January, the produce sections will contain few if any local products. In my opinion, this is why many people are skeptical, and even resistant, about the role of local foods in the food system. If local farms cannot supply their communities with food through the winter, then local products are relegated to the role of seasonal treats rather than trustworthy staples. While I fully acknowledge the systemic issues that shape our food system today, I also see opportunities for local producers to shape perception and build consumer confidence in locally grown goods through wintertime storage. Winter storage also offers some lifestyle and income opportunities for vegetable farmers. Storage crops aren't just for the industrial mega-farms; farms of any size can grow, store, and sell veggies locally, and turn a profit.

How is farming for winter storage different than growing veggies for summer markets? One huge difference is the distribution of work throughout the year. Rather than harvesting, processing, and selling crops throughout the growing season, storage farms have a concentrated harvest. The work of packing and selling crops moves primarily to the offseason, whereas the growing season is for growing and tending crops. I farm storage crops exclusively, and my year looks very different from that of my farmer friends. When I talk to them in September, they're usually exhausted from the grind of growing and selling crops all summer, whereas my busiest time, harvest season, is just getting underway. In winter, they're enjoying the slow time to recoup and plan, while I steadily chip away at my stored inventory. I find that I'm rarely burnt out from farming, but I extend my work throughout the year rather than cramming it into a few short months.

Finances on a storage farm are also a little different. Income arrives during the offseason rather than the growing season, which changes the financial flow of the year. It's certainly true that farming for winter storage comes with more overhead expenses. It costs money to build and purchase spaces for wintertime storage, climate-control systems, storage containers, and handling equipment. That said, these costs enable farmers to sell into a hungry and relatively open marketplace. Other storage farmers and I have found that the benefits outweigh the costs. The large wintertime demand also lowers the bar for marketing and advertising and makes it possible to get better prices for your produce.

The Finances of Winter Storage

You're probably wondering if winter storage works financially. In other words, can farming for storage not only turn a profit, but provide livable wages for both farmers and employees? Anecdotally, I can answer yes. There are many farms experiencing financial success with winter storage, and I profile a handful of them in part 4 of this book. Some of these farms use storage crops as their primary moneymakers, while others use winter storage as a side enterprise to supplement their other farming income. Both approaches can work.

Two factors in particular intimidate many farmers curious about winter storage: high initial costs for storage infrastructure and relatively low prices for most storage crops. These are justifiable concerns! Storage spaces, tools, and equipment *are* expensive, and no one is making six figures per acre from storage crops. That said, storage farmers can succeed financially so long as the infrastructure matches the scale of production.

Storage farms can work financially at scales both large and small, but every individual farm and situation will be unique. Annual income on storage farms depends on crop yields, sales outlets, and losses in storage. Yields also depend on many factors—soil fertility, climate, crop mix, and plant spacings, to name a few. In general, though, yields for mixed storage crops are usually between 10,000 and 30,000 pounds (4,500 to 13,600 kg) per acre. Farms on the lower end are usually larger and fully mechanized, whereas the highest yields often occur on small farms with intensive spacings. Four farms profiled in this book—ranging from 1 to 7 acres of storage crops—yield in the ballpark of 25,000 pounds per acre annually. Annual income from storage will vary according to the mix of crops and where they're sold. With yields of 25,000 pounds per acre, a farm might earn $37,500 per acre selling crops wholesale for $1.50 per pound. If they market those crops themselves—which takes both time and effort—that same farm might double their per-acre income to $75,000 if averaging $3.00 per pound. (This ignores storage losses, which can lower income by 10 to 15 percent.)

It depends on the farm, but many farms that finance their storage infrastructure pay somewhere between 5 and 20 percent of their gross income from storage crops toward their loan payments and annual operating expenses related to storage. Storage infrastructure includes things like storage spaces; containers; and harvest, washing, and handling equipment used specifically for storage crops. Operating expenses for storage

include obvious things like packing supplies, utilities, and labor (packing help), but also increased insurance and property tax costs due to the infrastructure. I'm not including operating costs related to growing and harvesting the storage crops because every farm incurs those expenses regardless of what they're growing. On my farm, expenses related to storage account for about 12 percent of my gross sales (which, for me, come entirely from storage). On other farms, it depends on how much infrastructure is financed; whether the farmer built new storage spaces or used existing buildings (and the same for equipment and building materials); and how the farmer's scale of production matches with their infrastructure and equipment. Also note that for many farms, storage crops aren't the only enterprise. If, for example, a farm can use some of its other infrastructure for storage-related tasks, the startup costs won't eat up as much of the income generated from storage.

You can use this ballpark range, though, to sanity-check your ideas for winter storage relative to your planned production. For example, if you project bringing in $100,000 annually from 2 acres of storage crops, costs related to the operation and financing of your storage infrastructure shouldn't exceed $20,000 annually (and preferably less). In this scenario, it would be wise to cap loan payments for storage infrastructure between $10,000 to $15,000 per year (10 to 15 percent) to leave room for storage-related operating expenses. With a loan that's at 5 percent APR on a 15-year term (typical for FSA loans), it might be ill-advised to finance more than $150,000 on storage infrastructure right away. If you can't keep your storage costs down, you might need to reevaluate how you sell your crops to generate more income. For example, you could adjust your mixture to include more high-value storage crops or try to sell a larger portion of your annual inventory directly to customers. You could also adjust your plans to make the storage infrastructure less expensive, such as finding used equipment or building some things yourself. These numbers may seem scary, but investments in storage infrastructure enable you to

Costs stemming from construction and operation of storage infrastructure on Offbeet Farm account for roughly 12 percent of annual gross income.

break into new markets and change the way you farm. Also know that costs vary widely; you don't necessarily need to build a state-of-the-art facility to get started with storage on your farm (see "Common Types of Modern Winter Storage," page 74).

Winter Sales

In most places, opportunities for winter sales will differ slightly from those available during the summer months. Even though winter farmers markets exist in some locations, many farmers markets shut down outside the growing season, forcing storage farmers to sell veggies elsewhere. Most storage farms sell their crops through wholesale and CSA, but some have access to farm stands, online sales, and limited special markets. Regardless of where you sell crops, you'll find that storage crops (by their nature) are more stable and keep longer than their summer cousins. Instead of packing orders days or hours ahead of time, you have the option to prep sales weeks before you actually need them.

Just like in summertime, many smaller farms choose to sell their winter produce directly to customers rather than through wholesale. Direct sales always bring more income, but you may be limited by how

In addition to wholesale distributors and a winter CSA, Boldly Grown Farm sells direct to customers through a farm stand, increasing overall winter sales. *Photo courtesy of Amy Frye.*

much you can sell. Some storage crops aren't sexy, and it can be difficult to make direct sales in large quantities. People don't often line up for turnips, and you'll be hard-pressed to make a living from kohlrabi if you need direct-market prices. I think CSA is a good model for ensuring sales numbers, but winter CSAs take some forethought to keep customers interested and engaged. There are fewer vegetable options than in summer CSAs, so excitement and variety must come from different colors, shapes, and flavors within a limited set of crops. The same reasoning applies to farmers markets, too: Unusual shapes and colors will draw customers to your stand better than a heap of beige potatoes (though I have nothing against beige potatoes). You'll find that some storage crops, though, like carrots and garlic, inherently generate excitement and drive sales.

Larger, more-mechanized farms can usually grow crops more efficiently and sell mostly at wholesale prices. While you'll take a hit on the price, wholesale provides opportunities for selling crops in much larger quantities. Most wholesalers, however, aren't interested in weird crops and varieties. They usually want staples—green cabbages, orange carrots, red beets, and so on—that most customers recognize and know how to use. That's not to say wholesalers won't take more-niche crops—sunchokes or black radishes, for example—but don't expect them to take these in huge quantities. You might have more luck selling quirky crops through direct marketing, where customers are more likely to expect the unexpected.

Setting Goals

Most successful farms set goals to work toward. Some goals steer a farm's long-term planning, such as saving money for a new piece of equipment or opening a new section of field. I encourage new farmers especially to use their personal preferences to steer their decisions regarding what they grow and sell. Don't like pruning tomatoes in a greenhouse? Great, you don't have to! Don't enjoy picking yellow leaves out of salad mixes? You don't have to do that either. Steer your production toward things you enjoy. That's one big reason why I'm a storage farmer; I genuinely enjoy growing and selling these crops!

Other goals are short-term, such as seasonal goals for income and production. I think annual sales and production goals in particular are important to running a successful storage farm. Income goals will help with covering the farm's regular expenses, paying back loans for storage infrastructure and equipment, covering personal expenses, and leaving extra for savings and long-term planning. Sales goals will help you meet obligations, such as fulfilling CSA shares or sales agreements with local stores. Whether it's an income or production target, these goals allow you to work out what's possible on your farm in terms of production, storage, and sales.

How do you create sales targets for storage crops? There are three elements to consider: amounts, locations, and timing. Put another way, how much of each crop are you selling, where are you selling them, and when will you sell them? Start your planning by making a list of your potential sales outlets. These might include winter farmers markets, groceries, restaurants, wholesalers, CSAs, online sales, or farm stands, to name a few. If you're homesteading, make a list of the crops and amounts you'd like to provide for yourself and your family. Your goal is to estimate the amount you can sell (or use) within each crop's storage life. (See part 3, the storage-crop compendium, for details on storage lives.) To do this, you need to estimate your sales rates. Get in contact with sales outlets you don't control—local groceries, for example—to get a sense for the amounts they normally sell. Some crops also sell differently depending on the time of year. One good example is pie pumpkins; sales peak before Thanksgiving in the United States and drop precipitously afterwards. For direct sales, try to determine a reasonable sales capacity for your customer base. Will your online store sell 100 pounds of carrots per week or 500 pounds? How many CSA customers do you have, and what's their monthly allotment of cabbages, potatoes, carrots, and so on?

Table 5.1. Estimating Seasonal Income

	Estimated Sales (Pounds)							Total Sales (lbs)	Price per lb	Income
	Oct	Nov	Dec	Jan	Feb	Mar	Apr			
Carrots								8,400		$20,500
CSA		500	500	500	500	500		2,500	$2.75	$6,875
Wholesale 1	200	400	500	500	400	400	300	2,700	$1.75	$4,725
Wholesale 2	100	100	100	100	100	100	100	700	$2.00	$1,400
Markets		500	1,500				500	2,500	$3.00	$7,500
Winter Squash								7,800		$15,025
CSA		1,000	800	800	800			3,400	$2.25	$7,650
Wholesale 1	500	1,000	500	300				2,300	$1.25	$2,875
Wholesale 2	200	400	200	200				1,000	$1.75	$1,750
Markets		400	700					1,100	$2.50	$2,750

Table 5.1 shows how a farm might plan their wintertime sales of carrots and winter squash. After listing out all their sales channels, they've estimated the sales (in pounds) through each channel for each month of the winter sales season. From these projections, they can estimate sales goals in terms of both weight and income. (Income estimates come from multiplying the overall sales weight by the price point for each channel.) This would be an opportune time to make adjustments if their income projections aren't meeting overarching income goals. However, before determining their harvest needs to meet these goals, they'll need to consider one more factor: crop losses in storage.

Losses in Storage

If you only harvest and store the exact amounts you plan to sell, you will almost certainly fall short of your sales goals. That's because losses due to spoilage and desiccation are inevitable parts of winter storage. Incorporating storage losses into your plans will help you manage risk and make your sales goals attainable. How much loss should you expect? Unfortunately, rates of loss can vary from crop to crop, farm to farm,

These beets were trimmed a little too closely and showed signs of rot after only a few months in storage.

and year to year. All sorts of things affect storage losses—management practices, local disease pressures, the weather, storage conditions, and even farmer error.

Estimating losses can be tricky, but my rule of thumb is to overestimate rather than underestimate. Your best guides will be records from previous years, but that requires both having a history on your farm and keeping records. Published results from storage trials are another option, but you must take these with a grain of salt. Because storage losses depend on so many factors specific to your farm, you can't expect your losses to exactly mirror those of trials conducted elsewhere. Rates of vegetable spoilage at grocery retailers are helpful guides for understanding the relative sensitivities of different crops. In retail terms, losses from spoilage are called *shrink*. Table 5.2 lists the shrink rates for common storage crops in US supermarkets. On the low end of storage-crop shrink are garlic, carrots, and potatoes, hovering around 6 or 7 percent. On the high end are pumpkins and winter squash, which have around 15 to 18 percent spoilage.[1] These numbers don't necessarily reflect loss rates in storage—supermarkets store and display vegetables in less-than-ideal conditions—but they give you a general idea of what to expect.

Amounts of spoilage also depend on the duration of storage and the rates of sales. At the beginning of the storage period, spoilage is usually low, with most losses caused by unnoticed damage or disease present at harvest. The longer you store a crop, the more losses you should expect. The rates of loss often increase dramatically toward the end of a crop's storage life. Table 5.3 shows storage losses—including water loss, decay, sprouting, and other issues that make a portion of the crop unsalable—for several common crops from published studies and trials. For example, a multiyear trial of storage onions in Finland—with onions harvested in early September and kept near 32°F—saw less than 1 percent loss after four months, 4 percent after six months, and then 54 percent after eight months.[2] Most biennial storage crops follow a similar pattern in storage, sustaining relatively low

losses until they near the end of their storage lives.[3] Winter squashes and pumpkins are annuals and, with few exceptions, don't display the same fortitude in storage. A trial conducted at Oregon State University found that multiple varieties of kabocha and buttercup squashes (*Cucurbita maxima*) experienced 11 percent losses after two months, 49 percent after four months, and 71 percent after six months in storage.[4] Notice that spoilage still increased over time, but losses progressed more steadily than in the onion trial.

How much loss in storage is acceptable, and how long should you plan to store and sell each of your crops? Ben Hartman, author of *The Lean Farm*, might say that all losses in storage are type-2 *muda*, or pure waste.[5] However, you'll need to accept that losses are an inevitable part of long-term storage. The trick is planning; don't let losses undermine your ability to run a profitable farm. Try to sell your stored crops before anticipated losses get too high, and remember that your

Table 5.2. Vegetable Shrink Estimates from Supermarkets

Vegetable	Average Losses (%)*
Brussels Sprouts	12
Carrots	6
Cabbage	11
Garlic	6
Kale	33
Onions	8
Potatoes	7
Pumpkins	15
Squash	18

* Averages over four years: 2005–6 and 2011–12.

Source: Jean C. Buzby et al., *Updated Supermarket Shrink Estimates for Fresh Foods and Their Implications for ERS Loss-Adjusted Food Availability Data* (USDA, Economic Research Service, EIB-155, 2016), www.ers.usda.gov /publications/eib-economic-information-bulletin/eib155.

Table 5.3. Published Storage Losses (as percentage of original harvest weight)

					Months in Storage		
	2	3	4	5	6	7	8
Beets[a]			13		57		
Cabbage[b]				23–45	24		
Carrots[c]			13		20, 14–35	23	
Onions[d]			< 1		4		54
Winter squash (*C. maxima*)[e]	11, 18		49		71		

Note: Numbers separated by commas indicate separate study results. All studies indicated that crops were stored within ideal conditions. The geographic locations varied widely, and results capture one or several years of trials.

[a] Renee Prasad and Susan Smith, "Red Beet Varieties for Storage in Lower Mainland BC," University of the Fraser Valley and BC Ministry of Agriculture, 2016, https://bcfoodweb.ca/sites/default/files/res_files/BeetStorageFactSheet.pdf.

[b] J. A. Cutcliffe, "Effects of Added Limestone and Potassium on Yield and Storage Losses of Cabbage," *Canadian Journal of Plant Science* 64, no. 2 (April 1984): 395–99; Christy Hoepting and Katie Klotzbach, *Final Report: 2009–2010 Storage Cabbage Variety Evaluation* (Cornell Cooperative Extension Vegetable Program, 2010), https://rvpadmin.cce.cornell.edu/uploads/doc_29.pdf.

[c] Ruth Hazzard, "2011–2012 Storage Carrot Trials," UMass Extension Vegetable Program, https://ag.umass.edu/vegetable/resources/winter-production-storage/storage/2011-2012-storage-carrot-trials; Terhi Suojala, "Effect of Harvest Time on the Storage Performance of Carrot," *Journal of Horticultural Science and Biotechnology* 74, no. 4 (1999): 484–92, https://doi.org/10.1080/14620316.1999.11511141.

[d] Terhi Suojala, "Effect of Harvest Time on Storage Loss and Sprouting in Onion," *Agricultural and Food Science* 10, no. 4 (2001): 323–33, https://doi.org/10.23986/afsci.5704.

[e] Ethan Grundberg, "2019 Kaboch Squash Variety Trial," Cornell Cooperative Extension, 2019, https://rvpadmin.cce.cornell.edu/uploads/doc_865.pdf; Jennifer D. Wetzel, "Winter Squash: Production and Storage of a Late Winter Local Food," (Master's thesis, Oregon State University, 2018), https://ir.library.oregonstate.edu/concern/graduate_thesis_or_dissertations/1j92gd35b.

inventory decreases as you sell your products. Thus, the higher rates of loss seen late into storage will affect a smaller proportion of your inventory. Here's another perspective: Göran Johanson of Goranson Farm in Maine explained that they begin triage-selling a crop when they've lost roughly 25 percent during storage.

I tend to plan conservatively and account for relatively high rates of spoilage. I expect the fewest losses from my root crops—such as carrots, beets, and parsnips—and usually account for 10 to 15 percent losses over six months. I use similar numbers for brassica crops like cabbages, Brussels sprouts, and kohlrabi to account for weight losses from removing aged wrapper leaves and desiccation. For onions, I plan more conservatively and expect higher rates of spoilage, usually 15 to 20 percent total. Weather around harvest

time can make proper harvesting and curing of onions tricky, so I like to give myself extra wiggle room for unexpected problems. Garlic is less finicky than onions, so I expect fewer losses in storage. When I store potatoes, I usually experience losses well under 10 percent. I expect the highest rates of loss to come from winter squash and pumpkins, and I plan to lose 25 percent of my initial harvest weight within about four months. Conservative planning makes it much more likely that you'll have a surplus rather than a shortfall at the end of the season, and you should plan for selling these extras as well.

Once you've estimated storage losses for each crop, you can calculate the amounts you need to harvest to meet your sales goals. Using proportions rather than percentages, divide the target sales weight by one

Squash Loss Scenarios

Imagine you harvest 10,000 pounds of kabocha squash at the end of September, and you expect them to last for four months in storage. As such, you make a goal of selling your entire inventory by the end of January but neglect to account for losses in storage. In one scenario, you plan to sell squash evenly over the four months, selling 2,500 pounds per month. In a second scenario, you plan to sell more squash up front: 4,000 pounds in October, 3,000 pounds in November, 2,000 pounds in December, and 1,000 pounds in January. If the kabocha squash actually spoil near the same rates observed in the Oregon State University trials—6 percent after one month, 11 percent after two months, 30 percent after three months, 49 percent after four months, and 60 percent after five months—what happens? In the first scenario, you're nearly out of squash by the end of December, carrying only 50 pounds into January. Total losses amount to 2,480 pounds, or nearly 25 percent of the original crop weight. In the second scenario, you run out in December, but you manage to sell more of your original inventory, losing only 1,740 pounds, or roughly 17 percent of the original crop weight. (For calculations, I multiplied spoilage rates by the remaining inventories to get the losses for each month.)

In the second scenario, you're able to sell roughly 700 pounds more squash than in the first scenario. Depending on your prices, that might account for more than $1,000 of real income. Thus, stretching a sales window toward the limit of a crop's storage life has real costs. Losses are an inevitable part of long-term storage, and you need to plan for them, but when expected losses are high, it behooves you to sell your inventory more quickly. However, this means a shorter sales window for your customers. If one of your goals is to provide food for your community deep into the winter, then you'll need to consider the trade-offs.

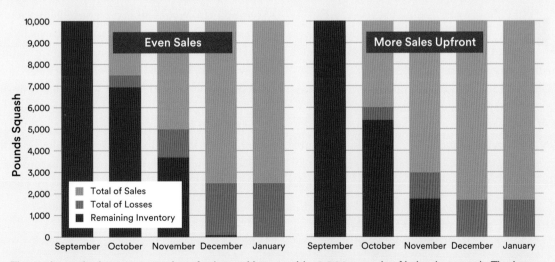

These charts depict two scenarios of sales and losses with 10,000 pounds of kabocha squash. The bars on the charts are cumulative, always accounting for the entire harvest. In scenario one (left), squash is sold evenly over four months. In scenario two (right), more squash is sold up front.

Losses are an inevitable part of storage. I keep a bin handy while packing orders to toss any spoiled pieces or trimmings, and these go to feed the pigs on another local farm.

pounds (8,400 pounds ÷ (1 − 0.10) = 9,333 pounds). In other words, they expect 8,400 pounds (90 percent) of the original harvest weight to remain salable while losing 933 pounds (10 percent) in storage. From there, they can allocate field space for the coming season by dividing harvest needs by expected yields, which ideally come from past harvest records. Since their past yields of carrots average about 35,000 pounds per acre, they should devote 0.27 acres to carrots to meet their winter sales goals.

Record Keeping

Before I became a farmer, I was trained as a scientist, and one thing I've carried forward from that time is a reverence for data and its role in decision-making. Running a farm business is complicated. There are a million things we *could* do, and many fewer things we *should* do to run our farms successfully. I try to keep excellent records and use that data to make informed decisions from year to year.

On a storage farm, I keep most of the same records that I would on any vegetable farm. Planting dates and locations, fertilizer applications, dates for thinning, and so on, are all things that I track. For the purpose of storage, I also record all harvest weights by crop and variety, and I maintain sales records that include the weights of each crop sold. These last two are useful for a few reasons. First, an important part of meeting sales obligations with storage crops is knowing your starting inventory. That tells you what's available and how to allocate it between your various sales outlets. For example, I know how many pounds (including expected losses) of a given crop are earmarked for the CSA. The CSA is my only obligation (because

minus the predicted loss to calculate the weight you'll actually need to harvest.

Table 5.4 returns to the example of the farm planning out winter sales of carrots and winter squash (see table 5.1, page 66). To meet their sales goals—8,400 pounds of carrots and 7,400 pounds of winter squash—they must harvest amounts that account for estimated storage losses. For carrots, this is 9,333

Table 5.4. Harvest and Field Space Needs

	Sales Goal (lbs)	Planned Losses (%)	Harvest Needs (lbs)	Past Yields	Field Space
Carrots	8,400	10	9,333	35,000/acre	0.27 acres
Winter Squash	7,400	25	9,866	25,000/acre	0.39 acres

customers pay upfront), and I can sell the excess inventory through a handful of markets and local groceries. I track those excess sales to determine when to stop selling to avoid undermining my CSA inventory.

You can also use harvest weights and sales records to calculate losses in storage. In this sense, losses account for everything you couldn't sell. While this doesn't separate moisture loss from spoilage from culls, it does provide useful information about the selling quality of each crop and the effectiveness of your sales strategies. I use these numbers to inform my future estimates for losses in storage. In the past, I've tried to measure my spoilage losses specifically but have found it impractical. When things go bad, they're often gooey and stinky, and I have no desire to handle them over a scale. To get more-detailed data about your storage losses, I suggest conducting trials on smaller samples of crops or specific varieties (see "Variety Trials for Storage," page 10).

Pricing, Timing, and Losses

One final note is that your pricing, sales rates, and losses from storage are interconnected, and they all affect your bottom line. The higher your prices, the slower your crops will sell, which will expose you to higher amounts of storage losses. Somewhere, there's a sweet spot that optimizes your income from storage, but you should take time to consider your ethics and goals for storage while making decisions. Maybe you're interested in maximizing your

> ## How Records Inform Sales
>
> If I need 1,500 pounds of parsnips for the CSA and account for 10 percent losses, I need to set aside 1,666 pounds (1,500 pounds ÷ (1 − 0.1) = 1,666 pounds). Let's say I harvest and store 2,200 pounds; then I have 534 pounds to sell as I wish. When my parsnip sales through markets and wholesale reach 534 pounds, I stop selling even if there are plenty of parsnips left. If any parsnips remain *after* the CSA (because I try to overestimate losses), then I can sell off whatever remains.

farm income. In that case, selling crops from storage as fast as possible to minimize losses is an effective strategy. Or perhaps you have a goal of providing your community with food as long into winter as possible. Maybe you accept higher storage losses as a service to your community, or maybe you charge higher prices to account for the costs of lost income. I don't have the answers. This is something you'll need to think about while planning for each season. It's important to generate enough income to continue farming (if that's the goal), but you should also let your other goals and values guide your decision-making.

This traditional, passive root cellar at a nearby Fairbanks farm is entirely underground and relies on soil to stay near the right temperatures. *Photo courtesy of Maggie Hallam.*

Chapter 6

Not Your Grandmother's Root Cellar

I have to give credit where credit is due; the title of this chapter comes from a presentation by Janaki Fisher-Merritt that I attended at the Midwest Organic and Sustainable Education Service (now Marbleseed) conference. In that presentation, Janaki described the evolution of the root cellars at Food Farm in Wrenshall, Minnesota (see "Food Farm," page 219). The title highlighted a central point of the talk—that modern commercial winter storage spaces are more sophisticated than traditional root cellars of old. That's not to say that traditional root cellars are obsolete. People have used root cellaring techniques for millennia to feed themselves through winter. Today, though, the demands on commercial growers are much different than they were even one hundred years ago. To stand out or even compete in today's markets, farmers need to meet higher standards established by the modern food industry, and to do that, we often need our storage spaces to go beyond what passive root cellars can deliver. Modern root cellars leverage traditional design and contemporary materials and technology to outperform what's possible with passive means alone.

Traditional Root Cellars and How They Fall Short

Before widespread use of electricity, people relied upon passive means to store their fruits and vegetables through the winter. Most often this involved using pits, holes, trenches, basements, and particularly root cellars, all of which rely upon the insulating properties of soil and the relative stability of deep-ground temperatures. With enough depth and insulation, passive storage spaces can usually maintain good storage conditions in a range of climates. While some are fully enclosed by wood or concrete, some are completely open to the soil, and most are not water-tight. Exposure to moist soil or water seeping through floors and walls keeps the humidity in these spaces high, which usually benefits the veggies. Most traditional root cellars are designed for passive ventilation, usually with intake and exhaust pipes positioned to promote natural air movement through convection. However, there are a few issues that often make passive storage spaces impractical for commercial growers and people wishing to maximize the storage lives of their vegetables during winter.

First, root cellars can be cramped and difficult to access. By necessity, root cellars are built into the ground, and there are usually stairs, ladders, or small doors to contend with. While small-scale home gardeners might not mind, commercial growers will likely find loading and unloading cramped root cellars awkward—if the veggies even fit. Root cellars built into hillsides or mounded with soil are more practical for large operations because they're accessible from ground level, where materials handling equipment can roll directly inside. These are relatively rare, though—often relics from older farming operations—and they suffer from other issues common among traditional cellars.

Temperatures in traditional root cellars are rarely optimal for storing vegetables in the best condition. Most are too warm for roots and other cold-storage crops but too cold for crops like squash and sweet potatoes that prefer warmer storage temperatures. (This doesn't hold in the American South, where deep soil temperatures are near-optimal for storing sweet potatoes, hence the long history of growing and storing them there.) If cellars do reach optimal storage temperatures, it might be late fall or early winter before they're ready to hold veggies. Again, this might not be a problem for small-scale growers with small gardens that can be harvested quickly. They can usually wait for their cellars to cool sufficiently, then nimbly harvest when the need arises. Commercial farmers, however, need more time to bring in large harvests and often can't wait for passive cellars to cool naturally. As such, they need other means of cooling crops immediately following harvest. Michigan State University's North Farm, for example, uses a passively cooled cellar for storing winter crops, but they also use a refrigerated shipping container to store crops in the right conditions while waiting for the cellar to cool down each fall (see "North Farm," page 177).

You might be thinking, why not just climate-control a traditional root cellar? While this can work in certain cases, many root cellars aren't designed for climate control; attempting to control temperature might be impossible or, at the very least, prohibitively expensive. The troubles are with thermal mass and insulation. (Thermal mass refers to a material's ability to absorb, store, and release heat.) Air has a relatively low thermal mass, and it therefore takes relatively little energy to change its temperature. Soil and concrete, on the other hand, both have lots of thermal mass, and it takes lots of energy to change their temperatures. Put another way, soil and concrete can store

substantial amounts of heat. While this is great for maintaining steady temperatures in passive cellars, those huge thermal masses will fight any attempts to change the natural air temperature. If you lower the air temperature in the cellar below the surrounding soil temps, heat will flow from the large pool in the soil into the air. This is precisely why North Farm uses a shipping container for temporary storage rather than mechanically cooling their cellar each fall. The uninsulated concrete walls, ceiling, and floor of their cellar, plus the soil beyond, prevent North Farm staff from economically cooling the space. If you try to heat the air in a cellar above the surrounding soil temperature, the opposite will happen—the soil will absorb the heat you put into the air.

Insulation is one way to circumvent these problems. Insulation will slow the flow of heat between the air in the cellar and the soil beyond. While it's relatively easy to insulate walls and ceilings—though you'll often need to insulate inward and decrease your usable space—floors can be tricky (see "Foundations and Floors," page 85).

Common Types of Modern Winter Storage

Where do commercial storage farms keep their vegetables if not in traditional root cellars? That partly depends on the farm's size, crops, and local climate, but winter storage spaces are as diverse as the farms on which they're found. Farms often have several different storage spaces to accommodate the special temperature and humidity requirements of various crops. You'll often see a farm's history and the story of its growth reflected in its winter storage spaces. Some farms outgrow their original storage rooms, build larger ones, and repurpose the older structures for new or different crops. A farmer might jump into winter storage and retrofit existing structures. Other times, farmers call on their summertime coolers to pull double duty for wintertime storage. Structures used for winter storage fall into a few general categories, each with its own merits and drawbacks.

The North Farm root cellar has a lot of uninsulated concrete, which makes it difficult to mechanically cool in the fall. *Photo courtesy of Sarah Hayward.*

Shipping Containers

When I arrived in Alaska and began building Offbeet Farm, several people suggested that I bury a Conex (a metal shipping container commonly used on ships) and use it as my storage space. While I was loath to do this for several reasons, the idea has some merit. In fact, there's a nearby farm in Fairbanks—Spinach Creek Farm—that's locally famous for their carrots, and they use a buried Conex for carrot storage for two to three months each fall. Theirs is uninsulated but buried into a hillside to protect it from early frigid weather.

Most farms use insulated shipping containers originally meant for refrigerated or frozen goods. While dimensions can vary, they're usually 7 to 8 feet (2.1–2.4 m) in both height and width, which is wide enough to fit two standard pallets side by side. Insulated Conexes used on boats are typically 10, 20, or 40 feet (3.0, 6.1, or 12.2 m) in length, whereas those

pulled by semitrucks vary in length up to 53 feet (16.2 m). Insulation depends on the container, but it's common to see a few inches of polyurethane foam, which has an R-value of around R-5 to R-6 per inch. (For more on R-value, see "Insulation," page 81.) While some farms use the original refrigeration units—if they still work—many cut holes in the sides of the containers to install CoolBot systems and add electrical outlets for lights, fans, and humidifiers.

The great thing about shipping containers is that they're inexpensive and take relatively little modification to make them functional for winter storage. This can be a boon for new farmers with little capital or for farms that are expanding quickly. (Another nice thing about shipping containers is that they hold their value.) Some farmers leave containers on the original trailer frames, but others set them directly on the ground. They're made for sitting out in the elements,

Open Door Farm in North Carolina uses several retired refrigerated shipping containers as their primary storage spaces. *Photo courtesy of Jillian Mickens.*

so you don't need to build a structure with a roof to protect them. Because of how they're built and insulated, you don't need to worry much about vapor barriers and condensation on structural components. This means that shipping containers can easily function as both summertime and wintertime coolers. They're also designed for storing heavy objects and have floors made for heavy handling equipment such as pallet jacks. That all said, their long and narrow layouts can make storing multiple crops difficult; you'll likely need to give up floor space to accessibly store more than two or three crops inside a single container.

While shipping containers are one of the cheapest storage options to purchase and set up, they can be expensive to operate in harsh climates. Insulation inside their walls, floors, and ceilings rarely exceeds 2 to 3 inches (5–8 cm), which means R-values aren't more than R-12 to R-18. That said, some containers

designed for frozen foods are insulated up to about R-26. At Tipi Produce in southern Wisconsin, their shipping container is the last storage space filled and the first one emptied because it's the most expensive to operate (see "Tipi Produce," page 211). Conversely, Open Door Farm in North Carolina uses shipping containers exclusively for their winter storage needs (see "Open Door Farm," page 197).

Walk-in Coolers

Walk-in coolers are another popular choice for winter storage on commercial farms. One big reason is that many farms already have walk-in coolers for their summer production. They can also be relatively inexpensive, especially if you can obtain used cooler panels. While designs vary widely, including those with premanufactured panels and homemade versions, walk-in coolers are basically small, insulated

This unique and functional walk-in cooler at Cripple Creek Organics near Fairbanks, Alaska, is retired from an airport, where it housed and protected electrical equipment. *Photo courtesy of Maggie Hallam.*

stand-alone rooms designed to hold a set temperature. They are often kept inside other buildings; many farms have walk-in coolers inside their pack sheds to store and process produce in one location. Walk-in coolers can also be stand-alone outdoor structures built to handle the elements. They don't have to be made from specially built cooler panels. I've seen plenty of DIY versions ranging from framed walls with foam insulation to a repurposed instrument storage shed from an airport. The through line is that they're large enough to walk in to, hence their name. Construction and installation costs can vary widely depending on the cooler's size, along with where and how you build it.

One strength of walk-in coolers is their versatility. You can build to whatever dimensions you need in whatever spot that fits, and working with prefab cooler panels is a lot like playing with Legos (albeit heavier and less colorful). Many prefab cooler panels are insulated to between R-20 and R-25, sometimes more, so walk-in coolers are generally less expensive to operate than shipping containers. They're also compatible with different cooling equipment, from CoolBot systems to commercial refrigeration units.

A downside of walk-in coolers is that without special construction, their floors usually cannot support rolling handling equipment like pallet jacks, stackers, and forklifts. The prefab floor panels aren't designed to handle the pressures exerted by fully loaded lifters on wheels, and you'll find similar problems with DIY insulated plywood flooring. In such cases, you'll be limited to using smaller carts and loading/unloading crops by hand. If you try to roll heavier objects—such as bulk pallet bins on pallet jacks—over plywood, you'll likely compress and damage the flooring and the insulation beneath (though I know farmers who do it anyway). In order to use large, wheeled loading equipment effectively, your walk-in cooler needs an insulated concrete slab. Condensation is another problem common to walk-in coolers used for both summer and winter storage (see "Vapor Barriers and Preventing Condensation," page 94).

Aboveground Insulated Buildings

Well-insulated aboveground buildings are completely viable options for winter storage, especially where winters are mild, but they're not common on commercial farms. I differentiate these from walk-in coolers in that storage rooms are integral to the structure of aboveground buildings, whereas walk-ins may be built inside but are separate from a building's structural components. Dedicated storage rooms and buildings are more efficient with precious floor space (since walk-in coolers should have some airspace around all four walls).

Whether building a new structure or retrofitting an existing one, designing and building an integral storage space has to be intentional from the start, and you'll need to take special precautions to avoid damaging the building's structural components with moisture (more on that in the next chapter). One popular and relatively inexpensive option is walling in a corner of an existing building. Putting up a new building will almost always be more expensive than shipping containers or walk-ins, but there can be huge benefits to doing so. First, you get to design and build whatever you want within your budget. This can include heated indoor space for washing and packing veggies, office space, and space for machinery storage and repair. You'll find it all invaluable! A building, so long as it meets local codes, will add value to your farm and property in ways the previous structures may not. While they aren't as energy-efficient as belowground structures, aboveground storage spaces are much cheaper and easier to build than their belowground counterparts.

Belowground Structures

Belowground structures are the gold standard for commercial winter veggie storage. This is partially because the legacy of traditional root cellars runs strong in our cultural subconscious, but also because soil does remarkable things for a building's thermal performance. While many people, myself included, may talk casually about soil's insulating properties,

The storage rooms at West Farm are integrated into the structure of the aboveground barn. *Photo courtesy of William Shinn.*

soil actually makes for lousy insulation. It's bad at slowing the transfer of heat from one place to another, especially if it's wet. The R-value of soil is only between R-0.25 and R-1.0 per inch. Instead, soil has high thermal mass, meaning it takes a lot of energy input or loss to change its temperature. This is why soil temperatures lag behind rising air temperatures each spring. No matter the time of year, soil temperatures tend to hold steadier than air temperatures, and this becomes truer at depth. The deeper you go, the closer soil temperatures are to being constant. How deep do you have to go before soil temperatures stop changing seasonally? It varies by soil type and climate, but it's usually between 12 and 40 feet (about 4–12 m), and those deep soil temperatures will be close to mean annual air temperatures for that location.[1] Temperatures will fluctuate further and further from the annual mean the closer you get to the surface.

Belowground structures are usually dug into hillsides or backfilled with soil mounded around them. This makes them accessible from ground level without ramps or stairs. Although they can be entirely belowground, they rarely are. Construction of belowground buildings is inherently more difficult and expensive because of the required earthmoving. However you do it, you need to dig a hole (or make a large pile if mounding around the building on flat ground). The building also needs to be strong enough to support the soil pressing on its walls, and resistant to moisture. Just because you're building belowground doesn't mean you get to skip insulation, either. While there can be specific circumstances in which you might omit insulation, insulating will improve your building's performance in most cases.

Belowground structures function well as storage spaces in both the winter and the summer because soil temperature fluctuations are less extreme than those for outdoor air. It takes less energy input to keep a belowground storage room at the right temps. Imagine it's early January, and a cold snap has outdoor temps hovering around 0°F. Depending on where you are, the soil at 3 feet below the surface might be in the mid-30s °F. It

I built the storage building at Offbeet Farm into the hillside, but my budget limited how deep I could go.

takes less energy to heat a squash storage room to 50°F when it is surrounded by 35°F soil rather than 0°F air. Fast forward to July, when you use the same below-ground room to store tomatoes at 50°F. Imagine a heat wave has outdoor temperatures near 90°F, but soil temperatures 3 feet down are in the mid-50s °F. Again, it takes less energy to cool that room to 50°F than something above ground.. The moderating properties of soil make the room more efficient and less expensive to operate in either case.

Geography and Costs

While meeting and interviewing storage farmers from around the United States and Canada, I noticed a geographic pattern regarding storage spaces. The southern farmers with milder winters were more likely to use less costly storage spaces such as shipping containers and walk-in coolers compared to their northern neighbors. There are definitely exceptions

out there, but I think the general pattern stems from the differences in upfront versus ongoing costs in different locations. Ignoring high material and labor costs in very remote locations, the cost of building stays roughly the same as you travel from north to south. What changes are the costs of cooling and heating the storage structures to maintain proper conditions. Most storage crops can remain in the field until cold weather forces us to bring them inside, and it's the severity of cold weather that follows that determines the ongoing cooling and heating needs in storage. Harsher winters necessitate more protection against the cold. Where winter weather is mild, it might be hard to justify a more expensive storage structure with lots of insulation when it's relatively inexpensive to operate cheaper facilities. Likewise, untenably high operating costs likely discourage farmers from using cheaper, poorly insulated structures in places with long and severely cold winters.

Chapter 7

Planning a Modern Root Cellar

Whether you're a gardener keeping food for your family or you're a large commercial grower, you can design and build a modern winter storage space that meets your needs and gives your crops the best chances in storage. This section isn't meant to be a construction guide, but instead provide an overview of materials and designs that lend themselves well to veggie storage. While some of these will apply to all common storage structures, including shipping containers and prefab walk-in coolers, most apply primarily to buildings and DIY walk-in coolers. My goals in this chapter are to give you useful information to design an efficient and functional space and to give you the knowledge to effectively communicate your needs to a contractor if you choose to hire one.

Insulation

If I had to pick the single most important component in winter storage spaces, I would choose insulation, hands down. Insulation is what protects your veggies from the harsh environment outside. Insulation is an expensive building material, but it's an investment that pays dividends down the road. Do a quality job insulating, and your storage structure will work more effectively and be less expensive to operate. However, you need to install the right types of insulation in the right places, or all the expense and effort might be for naught. Sources vary on the recommended amount of insulation, but they usually suggest somewhere between R-20 and R-30 for walls and ceilings.

Personally, I recommend more insulation in the ceilings—more than R-30 if possible.

Foam is one popular option for insulation. It is relatively light and easy to handle, and it has high R-values per inch relative to other forms of insulation. Foam is also rigid once hardened, making it easy to cut into any shape you need. Construction supply stores usually sell foam boards in 4-foot by 8-foot sheets of varying thickness. They're great for covering large flat surfaces like walls, floors, and ceilings. Some types are suitable as underground (below-grade) insulation that's directly in contact with soil and moisture, while others must be used aboveground. Generally speaking, foam board is not great for installing between wall studs because of the time and effort it takes to cut them. You also need to be thoughtful when installing them over interior or exterior walls, because they can trap moisture within a wall (more on that later).

The three common types of foam board sold at construction supply stores—EPS, XPS, and polyisocyanurate (commonly called polyiso)—are not interchangeable. EPS (expanded polystyrene) sheets are usually the least expensive, but they have the lowest advertised R-values—usually around R-4 per inch. The boards consist of small polystyrene beads that have been expanded with air and fused together with heat. The expanded polystyrene beads are "closed-cell," meaning they're impermeable to air and water, but the spaces between the beads are not. This means EPS foam can absorb water relatively quickly, but the water can drain away just as rapidly. Standard EPS has

R-Value

The purpose of insulation is to slow down the transfer of heat from one place to another. R-value quantifies that, allowing us to compare different types and amounts of insulation. At its core, R-value describes how a material *resists* the flow of heat—it is the imperial unit for thermal resistance. I find thermal conductivity, the inverse of thermal resistance, more helpful for understanding what R-values actually represent. Thermal conductivity describes how well a material conducts heat from one place to another. The imperial unit for thermal conductivity is BTU / (hr × ft^2 × °F), which is the number of BTUs transferred per hour for every square foot of surface area and degree of temperature difference in °F across the material. R-value is essentially the opposite of that, with its units flipped within the fraction. The building industry usually presents R-values accompanied by the letter R, such as R-15. For insulation, the higher the R-value, the more it resists the flow of heat. Generally speaking, the thicker a material is, the slower heat moves through it and the higher its R-value. If more than one material comprises a wall, their R-values are additive in cross-section. (For example, you would add the R-values of the drywall, insulation, sheathing, and siding to get the overall R-value for a standard wall.) If different parts of a wall have different R-values—because of a window, for example—their R-values are proportioned based on the surface areas of each part.

types are available. Historically, EPS was also more environmentally friendly to manufacture than XPS.

XPS (extruded polystyrene) sheets are more expensive than comparable EPS sheets, but they have higher advertised R-values, usually around R-5 per inch. (That R-value can drift downward over the years, however, as internal gases of blowing agents leak out.) XPS sheets are the colorful ones; they're often blue, pink, or green, depending on which company manufactured them. They're made by mixing liquid polystyrene with blowing agents—which make the foam expand during drying—and extruding the mixture into flat sheets. XPS is a closed-cell foam, meaning it's largely impermeable to air and water, but the sheets will absorb moisture slowly over time. Unlike EPS, water-logged XPS sheets take a long time to dry once wet and often stay permanently wet when used below-grade, which lowers their effective R-value by about 15 percent.[1] Standard XPS sheets are stronger than EPS and are usually rated to compress only 10 percent of their original thickness at 25 psi (good enough to go under most concrete slabs). Recent changes to XPS manufacturing have made it a more environmentally friendly choice than it was in the past.

The last type of foam board you'll commonly see at hardware stores is polyiso. These panels are made by mixing a few chemicals with a blowing agent and pouring the mixture over a foil or paper backing. The chemicals react to generate heat and the gases expand to form a foam. I don't recommend using polyiso panels with cold-storage structures for a couple reasons. First, they're more expensive than the other foam boards. You may think that's because they have a higher R-value per inch. While this is true in some circumstances, polyiso panels have a big problem with thermal drift, meaning they lose insulative value over time. (This occurs because the blowing agent, which is a better insulator than air, leaks out from the foam and is replaced by air. This also happens with XPS to some extent.) They also lose R-value as the temperature drops. At temperatures common to cold storage, the effective R-value of new polyiso panels is probably

lower compressive strength than standard XPS, so it might not be suitable under all concrete slabs or footings, but more expensive, denser versions of both between 4 and 5 compared to their advertised resistance of R-6, which negates their relatively high price tag.[2] Polyiso panels are also not rated for ground contact and should never be used outside a below-grade wall or under a slab.

Where and when should you use the different types of foam board? In my opinion, EPS is the most versatile and affordable option with the fewest downsides. EPS has a relatively low environmental impact and its R-value won't drift over time. Despite being relatively permeable to moisture, EPS is a good choice for belowground insulation because it also drains quickly when the moisture goes away. That said, XPS is always a sound choice for insulating under slabs because of its high compressive strength.

Spray polyurethane foam, or SPF for short, can be handy in new construction but is especially helpful during retrofits. Installing SPF over a wide area usually means hiring a contractor, but the process is fast, and the time savings might be worth the expense. As opposed to most other types of insulation, spray foam conforms to all sorts of nooks and crannies that are otherwise difficult to insulate. Closed-cell SPF is also rigid after curing and adds to the structural integrity of a building, which can be especially helpful when retrofitting older buildings. For smaller areas, you can use spray-foam products for consumers, such as Great Stuff made by DuPont. Although you can choose between open- and closed-cell foam, I recommend closed-cell because it won't absorb moisture in the humid environments used in winter veggie storage. SPF creates a vapor barrier wherever it's sprayed. Be careful, though—this characteristic might be convenient in some circumstances but lead to moisture issues in others. Most SPFs have an initial R-value over R-6, but that will decrease over time as blowing agents leak

Sheets of EPS foam board (furring method) add additional insulation to the ceiling of Offbeet Farm's storage building.

The potato storage room at Mythic Farm in Blue Mounds, Wisconsin, was retrofitted from a barn basement using SPF foam as insulation.

Mineral wool batts fill the stud cavities of the offset stud walls in the Offbeet Farm storage building. After two layers of 3.5-inch batts, the walls are insulated to R-30 with minimal thermal bridging.

from the foam. As with manufacturers of XPS, makers of SPF have recently made changes to their formulas to reduce the global-warming impacts of their products. If you're concerned, make sure the company you hire doesn't use HFCs as blowing agents in the foam.

Apart from foam insulation, there are also insulation batts, which can fit between wall studs and make layers over large areas, and loose-fill insulation commonly blown into attics and walls. Of the three common materials used—fiberglass, mineral wool, and cellulose—I highly recommend using mineral wool (also called rockwool) and avoiding both fiberglass and cellulose. The main problem with the latter two is that they absorb moisture and lose their insulating fluff when wet. If they remain wet for long periods, mold and mildew problems are likely to develop in the absence of chemical inhibitors. Mineral wool, on the other hand, is hydrophobic, meaning it repels water. It also receives a coating of oil during the manufacturing process that enhances water repellence. Mineral wool is made primarily from crushed rocks and slag, a by-product of steelmaking, which are melted and spun into dense fibers. You can buy it in batts pre-cut to fit common stud spacings (for example, 16 and 24 inches on-center) or as loose fill for blowing into attics and walls. In batt form, mineral wool usually has R-values between R-4 and R-4.5 per inch.

While it's often easier and cleaner to build with entirely new materials, sometimes used or repurposed insulation can be just as functional, reduce your project costs, and prevent good materials from going to the landfill. I think this is particularly true with rigid foam insulation. When building the winter storage building at Offbeet Farm, I found used pieces of XPS insulation that came from the roof of an old school, and I got them at a quarter of the price of new sheets. They weren't in pristine condition—many were sun-damaged and waterlogged—but they were still entirely useful for beneath the concrete slab. I also found someone selling bundles of EPS sheets that were unused and overordered from a project. These were around half the price of new sheets, and it was a win-win situation for both me and the seller. If you start asking around and exploring places where people sell unwanted items, you might be surprised by what you can find!

Foundations and Floors

Foundations are to buildings like directors are to films; they're, well, foundational. They hold things together and hold them up, though they're mostly hidden from view. I marveled at the time, effort, and expense that went into the foundation of the storage building at Offbeet Farm. Even though it's out of sight, the foundation is vitally important for the function and longevity of the building. In addition to supporting the building above, it's the basis of a square, plumb, and level structure. If the foundation is wonky, the building probably will be too.

Foundations for winter storage buildings have similar functions to those used in residential construction. They help the building maintain the right temperatures and support the weight of the building and its

contents without shifting or heaving. There are many types of foundations to choose from, but I don't recommend using crawl-space or pier/piling foundations unless you're not storing much weight or you absolutely have to (for example, if you're building over permafrost). Instead, I recommend building concrete foundations either on-grade (at ground level) or in the style of walkout basements. I do not recommend traditional basements because of the difficulty of getting veggies in and out. Although concrete slabs aren't entirely necessary, I think they're worth the extra effort and expense. Slabs are easy to insulate from below during construction, and they enable you to use heavy handling equipment with wheels. I cannot overstate how much I love using wheels!

On-grade foundations are easier and less expensive to build than walkout basement–style foundations, but they lack the benefits of surrounding soil (discussed below). Concrete on-grade foundations usually consist of footings—deep and wide concrete bases that support load-bearing walls or structural poles/beams—overlayed by concrete slabs. Footing size and depth depends on soil types, frost depth, and the weight they need to support. At minimum, footings need to have enough surface area to avoid warping the soil with the weight of the building and the load within (the weight of stored veggies). Soil capacity is the maximum weight that a soil can support before deforming, typically measured in pounds per square foot. Bedrock, gravel, and sandy soils can generally hold a lot of weight without deforming, while silt and clay soils easily deform and swell when saturated with water. Traditional building practice is to set footings below the local frost line, and depending on your soils, you may need a layer of gravel beneath the footings to remove moisture and spread weight. Local engineers and builders can tell you typical foundation specs for your region and soil types.

A popular style of on-grade foundation in northern climates is the frost-protected shallow foundation, or FPSF for short. In this design, a deep layer of gravel or crushed rock is placed beneath and around footings to

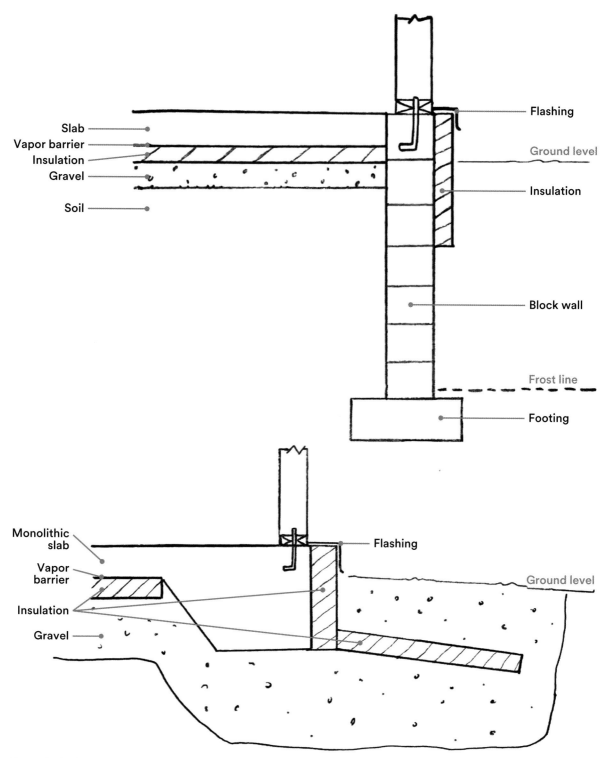

Slab

Vapor barrier

Insulation

Gravel

Soil

Flashing

Ground level

Insulation

Block wall

Frost line

Footing

Monolithic slab

Vapor barrier

Insulation

Gravel

Flashing

Ground level

Example designs for a standard foundation with footings below frost line (top) and a frost-protected shallow foundation (bottom).

drain away excess moisture, and insulation is placed vertically against the footing walls, horizontally into the soil like wings, and sometimes below the slab. FPSFs are popular choices because they circumvent building codes that require footings to be below the frost line. This can be quite deep in certain locations; the northern half of Minnesota, for example, requires standard footings to be no less than 5 feet (1.5 m) deep. Digging and pouring concrete that deep can be costly; FPSFs are a cost-saving alternative, which is great for farms on a budget. You can find specifications for FPSFs in regional building codes, and most use XPS foam as the vertical and horizontal below-grade insulation.

Below-grade foundations in the style of walkout basements are more complicated and expensive to build, but the soil surrounding them provides thermal benefits. Footing and slab designs can be similar to those used in on-grade construction, but perimeter footings are usually beefier to support the weight of foundation walls and additional stories when present. The big difference is the presence of foundation walls, which are often, but not always, made from either concrete blocks or poured concrete. The costs for each method vary widely depending on your location relative to a concrete plant, your skills in construction, the building site itself, and your willingness to schlep concrete by hand. The materials for cinderblock walls are less expensive, but it takes a lot of skill and time to build plumb and square walls. Personally, I would need to hire a mason. Block walls are prone to water infiltration and they're poor insulators. You'll need to put in extra time, effort, and expense both waterproofing and insulating the wall exteriors. Poured concrete walls, on the other hand, can be more expensive, but

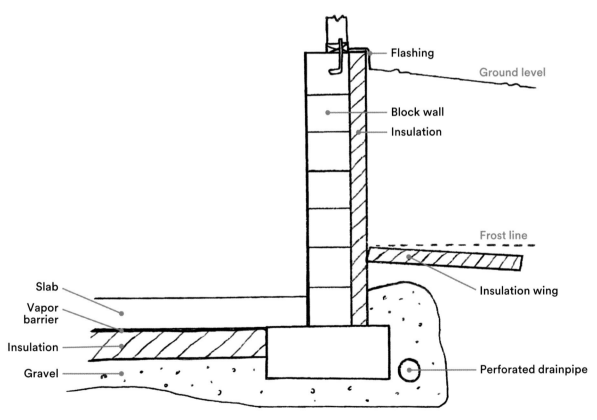

Example design for a basement foundation with a block wall. The basement wall could also be, for example, poured concrete or insulated concrete forms (ICFs).

The bottom half of the Offbeet Farm storage building has insulated concrete form (ICF) walls. The first course of ICF blocks was in place while the slab and footings were poured, while the other courses of ICF blocks (some visible in the background) were part of a later, second pour.

they are less time-consuming to build and require less work to waterproof. Poured walls are also stronger than block walls and can generally handle more lateral pressure from soil. Building concrete forms that are square, plumb, and level, and (importantly) won't collapse during the pour takes skills that some farmers don't have. That said, poured concrete walls generally take fewer skilled labor hours to assemble than block walls. Like block walls, poured-concrete walls are poor insulators, and you'll need to spend additional time and money insulating their exteriors.

While cinderblocks and traditional poured-concrete foundation walls are options—usually built or poured directly upon completed concrete footings—I'm an advocate for using insulated concrete forms, or ICFs for short, to build full-height foundation walls or even short stem walls. ICFs are made from EPS foam with plastic reinforcement and, when finished, create reinforced concrete sandwiched between two layers of foam. The amount of insulation varies, but the finished walls are often around R-25. A huge benefit of choosing ICF walls over both cement-block and traditional poured concrete is ease of installation. ICF blocks are easier for novices and professionals alike to assemble, reinforce, level, and brace. The blocks have an abundance of slots and hangers for placing rebar reinforcement, and they stack and assemble much like Legos. They have the same water-resisting properties as traditional poured walls, but it's a good idea to protect aboveground and belowground EPS foam with rubberized coverings, metal flashing, or siding materials. ICF blocks usually have plastic bands embedded in the foam that allow you to fasten wall covers—sheet metal, plywood, sheet rock, FRP panels, and so on—directly to the foam walls.

Not all basement or on-grade foundations have slabs, but concrete slabs make excellent floors for veggie storage. While it's common to pour the footings and slab separately, floating slabs are another option that can be simpler and less time consuming to build. The design is simple; the footings and slab are poured together, and the footings, if present, are just deeper sections of the slab around the perimeter of the building pad. Just like with footings, you may need to place gravel beneath the slab depending on your soils, and it's good practice in most circumstances to insulate the slab by pouring it over foam board. Slab thickness varies depending on the use. Most residential slabs, including driveways and standard garages, are 4 inches (10 cm) thick, while many warehouses have 6-inch (15 cm) slabs to accommodate heavy equipment. Depending on your building's design and layout, you can vary slab thickness in different sections to fit different uses.

In most cases, you should insulate below concrete slabs and around footings and foundation walls. Sometimes builders insulate below footings with high-pressure XPS foam, but this isn't a common practice. Insulating belowground walls and floors might seem counterintuitive for root cellars—you're trying to utilize all that ground heat, after all—but it really depends on the temperatures you'll be maintaining relative to your local deep soil temperatures. (As a reminder, deep soil temperatures will be close to the mean annual air temperature.) For example, if your cold-storage room will be 32°F (0°C) and deep soils are around 42°F (6°C), you'll definitely want to insulate the slab and walls. Otherwise, your refrigeration system will fight against the constant influx of ground heat.

This scenario happened in the storage building at Food Farm in Wrenshall, Minnesota. They intentionally left a portion of the cold-storage room slab uninsulated, thinking the influx of ground heat would save them from heating during the winter. Instead, the veggies provided more than enough heat through respiration,

Concrete Strength and Reinforcement

Concrete comes in different compressive strengths, which concrete companies rate in pounds per square inch (psi). This refers to the amount of compressive force concrete can take before it cracks. Companies usually sell concrete in the range of 3,000 to 5,000 psi. While 3,000 psi concrete is fine for most residential needs, you'll want to use at least 4,000 psi concrete for most footings and for slabs that will support heavy handling equipment such as pallet jacks and forklifts.

Concrete is impressively strong when compressed, but it's surprisingly weak when bent. Since concrete has low tensile, or bending, strength, you'll need to reinforce it with materials with high tensile strength, most often steel rebar. Although rebar comes in many sizes, the most common sizes used in foundations and slabs are #3, #4, and #5. These differ in thickness and are ⅜-inch, ½-inch, and ⅝-inch in diameter, respectively. Builders often place rebar in grid patterns in walls and slabs, while long, linear pieces are common in footings. They use pieces bent at 90-degree angles to go around corners and strengthen connections between footings, walls, and slabs. Rebar size and spacing largely determines the final tensile strength of the concrete. While #3 rebar is adequate for residential driveways and garages, you may want #4 or #5 in slabs used for veggie storage. Number 4 rebar is common for poured-concrete walls and concrete-block walls, while #5 rebar is common in footings. Rebar spacings depend on the design specs, but 16 inches is common.

and the extra ground heat wasn't helpful. They have to run their refrigeration system all winter to keep the room cold enough (see "Food Farm," page 219). The only situation in which to leave walls and floors uninsulated is when your desired room or building temperature approximately matches deep soil temperatures. Even then, you should insulate to at least the frost line to prevent issues with cold spots and to meet building codes. Many builders also install horizontal insulation "wings" just below the normal frost line to prevent deep frosts from damaging walls or the veggies within.

Installing thermal breaks in concrete slabs will also help you maintain the proper temperatures in your storage rooms and the like. Thermal breaks slow the transfer of heat from one section of concrete to another, and they're handy for isolating sections or rooms kept at different temperatures. You install them by embedding foam board spacers—usually XPS—before pouring a slab. (Sometimes, you'll need to support the foam board with lumber if you're only pouring from one side.) For example, if you have a concrete pad that extends outside the building, a thermal break will prevent heat from leaking out of the building via the slab during the winter or vice versa during summer. Likewise, thermal breaks between rooms set at different temperatures will prevent heat from leaking from one room to another through the concrete.

I've touched on moisture concerns already, but there are things you can do to prevent water from collecting around your foundation and damaging it. One of the most important things is site prep and grading. Whether it's you or someone you hire, make sure that the ground near the building pad is sloped to direct water away from rather than toward the building. This isn't always easy and might take some skilled work with heavy machinery. Another preventative measure is to put perforated pipe around the footing (below the level of the slab) and at least one pipe draining water away from the structure. This is usually done by placing perforated pipe inside a bubble of gravel that's wrapped in geotextile fabric. The fabric stops soil from

clogging the pipe over time, while the gravel helps drain away excess water. Installing rain gutters on the roof is another way to keep excess water from collecting around the foundation. Lastly, damp-proofing and waterproofing layers over concrete foundation walls will help prevent water infiltration via both wicking and seeping through cracks. Waterproofing over ICF walls will also improve the insulative value of the exterior foam and protect it over time. There are countless products and materials out there, including spray-on waterproofing and peel-and-stick products. Be sure to choose a product compatible with the surface, be that foam or concrete. It's also common practice to place polyethylene sheeting under footings to prevent them from wicking up soil moisture.

I highly recommend hiring an architect or an engineer to create and/or certify your building plans, and especially to assign specs for the foundation. Throughout this section I've been intentionally vague about thicknesses and placements of gravel, concrete, insulation, rebar, and the like because they're all context dependent, and every situation is different. Certified building plans will specify the types of concrete, gravel depths, insulation thicknesses, rebar sizes, rebar spacings, and more to ensure your building functions as it should and to keep the foundation from failing. While you can hire an engineer or an architect to create building plans from scratch, you can also bring them your own plans to certify. They will usually make new drawings based upon your plans, and the drawings will include detailed building specs to meet codes, structural requirements, and your needs for the building.

Walls and Ceilings

Just like foundation designs, designs for walls and ceilings in buildings used for winter veggie storage can be similar to those used in residential construction. However, these walls and ceilings need to maintain temperatures and humidities not regularly found in homes. Oftentimes you even need to confine a particular set of conditions to a single room or to multiple rooms. Wall and ceiling designs can incorporate standard framing techniques: studs stacked between base- and top-plates for walls and ceilings from the roof trusses or floor joists of a second story. Where things might deviate from usual building techniques is with vapor barriers (see page 94) and insulation. In particular, you'll want to prevent a phenomenon called thermal bridging. This occurs when the structural components of the building act as heat conduits, bypassing the insulation. The way you combat this is by adding thermal breaks—insulation or space that breaks thermal bridges between the two sides of a wall or ceiling.

I have a friend here in Fairbanks whose house was clearly built without thermal breaks in mind. Every time I visit during the winter, their home's wall studs are clearly visible through the frost patterns that form on the exterior siding. Whereas most of the house is covered in a thin layer of frost, there are frost-free parallel lines—I've never measured them, but I'd bet they're each an inch and a half thick—conspicuously spaced every 2 feet or so. Heat from inside the house is leaking either around or through the wall studs to melt the frost. If you used a thermal camera inside the house, wall studs, base-plates, and top-plates would be clearly visible as cold spots. This isn't a serious issue in most homes—although it definitely affects heating bills—but cold spots in winter veggie storage might lead to frozen produce or generate unwanted condensation.

You can't rely on wooden structural components to provide substantial insulation. Wood is a relatively poor insulator, only about R-1.4 per inch for softwoods. (That said, wood has high thermal mass

Patterns of frost formation reveal thermal bridging—in which heat leaks from inside via the wooden framing—at a nearby friend's home in Fairbanks. *Photo courtesy of Seth Adams.*

compared to most conventional insulation, which is why log homes can perform well in cold climates.) A 2 × 6 wall stud, for example, only provides an insulating value of R-7.7. Although omitting thermal breaks won't substantially hurt the overall R-value of a wall, it creates the potential for cold spots.

To avoid thermal bridging, no structural components should span the full width of a wall or ceiling. I suggest one of two techniques for creating thermal breaks. The first is a technique called double-wall construction. It's a common building design in interior Alaska to create super-insulated homes. True double

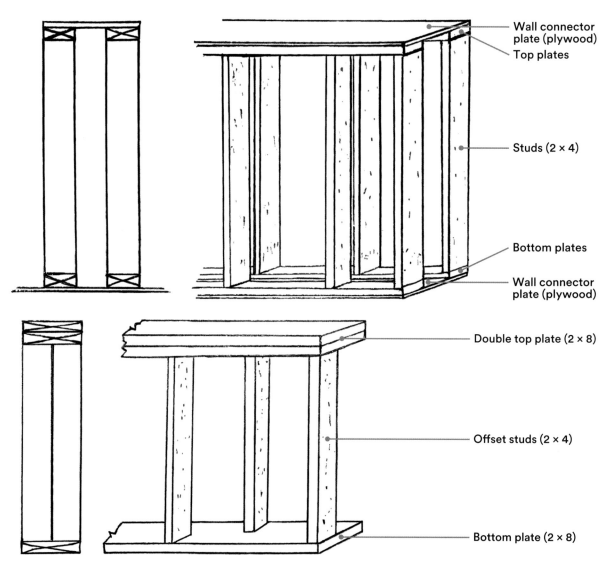

Two methods of building walls with thermal breaks: double walls (top) and offset studs (bottom). There are many techniques for building double walls, but this diagram depicts the separate 2 × 4 walls joined at the top and bottom with plywood connector plates. Insulation can fully fill the space, or there can be an air gap between the wall frames. The offset stud design shown here features 2 × 8 bottom and top plates with 2 × 4 studs. Insulation fully fills the cavities.

walls are just like they sound: two separately framed walls separated by either insulation or an air space. These are often two 2 × 4 wall-frames held in place on wide plywood base- and top-plates, leaving a space between the walls. If that space is dead air, the wall's R-value will be around R-31; if it's foam board, the wall may be upwards of R-40. Another, less expensive double-wall design is offset studs. One way to execute this design is with 2 × 4s as studs on 2 × 8 base- and top-plates. On each side of the wall, 2 × 4s sit at a fixed spacing, say 24 inches on-center, but they are offset from the studs on the opposite side. Regular 3.5-inch insulation batts can fit in the cavities on each side of the wall, but no wood spans the full wall width except at the base- and top-plates. In this configuration, walls are near R-30 with very little thermal bridging. This is the technique I used for framed exterior walls in the storage building at Offbeet Farm, and I found the materials were less expensive than traditionally framed 2 × 8 walls; two 2 × 4s were cheaper than one 2 × 8.

Another way to prevent thermal bridging in walls and ceilings is to cover one side of the framing with foam board. (I do not suggest covering both sides of a framed wall with foam board because of the dangers of double vapor barriers—more on that later.) There are two common ways of doing this: sheathing and furring. *Furring* means to fasten wooden "furring strips" directly over, and usually perpendicular to, wall framing, and to fit sheets of foam board between the strips. People usually leave either 2 or 4 feet between the furring strips to accommodate either half or full foam sheets. Fastening the final wall covering—sheet metal, for example—over the furring strips is relatively easy and doesn't require special long fasteners. (Here's a tip: Start from the bottom and work your way up, adding furring strips and foam board together. If you install all the strips first, you might find that foam sheets don't fit well between the spaces.) The furring method will eliminate most thermal bridging, except where the furring strips and studs overlap.

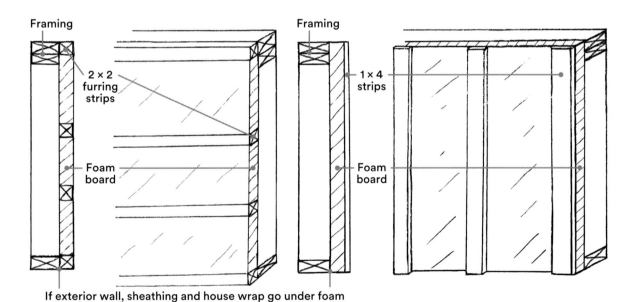

If exterior wall, sheathing and house wrap go under foam

Two more methods of building walls with thermal breaks: furring with foam board (left) and sheathing with foam board (right). The furring method places foam board between furring strips for easy fastening of top wall cover. The sheathing method uses long fasteners to attach strips over the foam board (to vertical studs). Exterior wall coverings attach over the strips. If it's an exterior wall, then plywood sheathing and house wrap would go directly onto wall studs underneath any foam board layers.

Sheathing means to fit the foam sheets tightly together with no gaps. This method does a better job of eliminating thermal bridges but makes fastening the final wall layer more complicated unless you're gluing the final layer in place, as is common with fiberglass reinforced plastic (FRP) panels. Many builders use 1 × 4 strips over the foam as fastening anchors for a top wall covering. The strips can either be perpendicular to or parallel with wall studs, and fasteners holding the strips should penetrate studs at least 1.5 inches.[3] Depending on the thickness of your foam board, you'll sometimes need unusually long fasteners for this job!

While either sheathing or furring with foam board, you'll have to decide whether to do this on the interior or exterior of the building. (You usually won't have a choice with ceilings, since adding foam to the inside is often the only viable option.) While insulating on the inside will reduce some usable volume, adding foam board to the exterior of a building may cause condensation issues inside walls by moving dew points onto structural components (see the next section, "Vapor Barriers and Preventing Condensation"). If you do add foam to the exterior, both the structural sheathing and house wrap should be installed under the foam boards.

Beyond regular framing techniques, there are wall and ceiling systems that inherently preclude thermal bridging. One popular style is the structural insulated panel, or SIP for short. These are basically a layer of foam—usually EPS, although some manufacturers will use other types of foam—sandwiched between OSB sheets. You can use SIPs for structural walls, ceilings, and even roofs. Compared to traditional framing, SIPs will cost more in materials, but construction is generally faster. I advise caution, however, for using SIPs to build high-humidity storage spaces. Without proper care during installation to prevent air leakage, SIP walls and roofs have been known to develop serious condensation problems at the joints between panels. Wet OSB tends to swell and rot, losing structural integrity over time. I highly recommend consulting with or hiring a contractor experienced

with SIPs if you plan to build with them. Compared to other building materials, the margins for error are small, and you don't want to face costly repairs only a few years after construction. There are also metal insulated panels, or MIPs for short. These tend to be very expensive and aren't much different from prefab cooler panels. They have the same labor-saving benefits of SIPs but without the potential for moisture damage when installed incorrectly.

Vapor Barriers and Preventing Condensation

I've alluded to the importance of vapor barriers several times, but what are they and why are they important for winter vegetable storage? Even the driest storage conditions are more humid than typical homes during the winter. Without special attention during construction to stopping air leakage and the diffusion of water vapor, our storage spaces may develop large amounts of condensation in places easily damaged by moisture.

To prevent condensation, you need to impede the movement of both air and water vapor with vapor barriers and vapor retarders. The goal is to prevent condensation from forming within walls, ceilings, or roofs. In the worst-case scenarios, prolonged wetting from condensation can cause a building's wooden structural components to rot. Condensation can encourage mold growth, render some types of insulation impotent, and cause short circuiting in electrical lines.

Condensation forms in walls whenever humid air contacts a surface below its dew point. Why does this happen? Whenever there are different temperatures on opposite sides of a wall, there is a temperature gradient across the wall. For example, if one side of a wall is at 32°F and the other is at 60°F (16°C), the temperature at different points inside the wall is somewhere in between and follows a steady, linear gradient from one side to the other. Different materials—such as plywood, insulation, or drywall—change temperature more gradually or more precipitously depending on their insulative value. The point at which humid air

The beginnings of a vapor barrier. Sealing the 6-mil polyethylene sheeting to the top plate with polyurethane caulk creates an air seal around the entire ceiling. Poly sheeting on the wall will be sealed (with caulk) directly onto the lip of sheeting overhanging from the ceiling to make a "perfect" seal.

drops below its dew point is where the condensation forms. This may be inside insulation, plywood sheathing, wall studs, cold electrical lines, or anywhere else humid air contacts cold surfaces. The purpose of vapor barriers is to limit both air movement and the diffusion of water vapor, thereby limiting the amount of condensation forming in troublesome places.

When adding vapor barriers to walls and ceilings, the basic rule of thumb is to place them on the warm side just under the final finishing layer (drywall, plywood, FRP panels, siding, and so on). This prevents the relatively warm and humid air from reaching dew point as it moves through the temperature gradient inside the walls. For winter storage spaces in cold climates (where most of the winter is spent below freezing), this means putting the vapor barriers inside.

In milder climates, and especially in humid ones, vapor barriers on the outside will prevent in-wall condensation from forming when warm outdoor air touches cold, indoor surfaces. If you're in doubt, you can use wall dew point calculators to try out proposed wall designs in your local climate. The University of Vermont Extension Agricultural Engineering program makes a good one (see resources, page 237). Try plugging in some monthly averages for temperature and humidity during the times your winter cooler will be running. The calculator will notify you if condensation occurs within the wall and which direction it's coming from.

But what if you plan to use the winter storage space as a summertime cooler as well? In many northern climates with cold winters and warm summers, this presents a problem. During the winter, humidity inside the cooler will try to escape through the walls, whereas in the summer, humidity from outside will try to creep in. This is because water vapor naturally diffuses from areas of higher concentration to lower concentration, and (relatively) warm air holds more moisture than cold air. Thus, veggie coolers in northern climates experience a *vapor drive*—where water vapor *wants* to move through walls and ceilings—that changes seasonally. If you design the vapor barrier to protect against winter conditions, condensation will likely form in the walls when you use the cooler during the summer, and vice versa.

So, what should you do? Put vapor barriers on both sides? I put this question to Chris Callahan, an extension associate professor of agricultural engineering at the University of Vermont. He said that it's a conundrum and that people disagree on the solution. One option, he said, is to create perfectly sealed walls and ceilings with air and vapor barriers on both sides. He stressed that anything less than perfect seals around the vapor barriers won't do because water vapor would then slip inside the walls and condense. Another option is to seal one side of the wall and provide ventilation on the other side to give moisture an escape route. In the best-case scenario, there would be a small air space spanning the wall with ventilation on the top and bottom to promote convective drying. In practice, I think the second option is more feasible. Creating a perfect vapor barrier is difficult—not only does the initial seal need to be flawless, but you also can't use regular fasteners to attach the finished wall material because fasteners will break the perfect seal. That perfect seal also needs to last through time as the building settles and materials age. Frankly, I think that relying upon perfection is a risky proposition. If there's any mistake, you'll have a situation in which moisture gets trapped inside a wall without a good way of getting out.

A wall dew point calculator can help you work through different scenarios if you decide to design a cooler for both summer and winter use with just one

Relative Humidity, Dew Point, and Temperature

Why does condensation form? It's all about the relationship between relative humidity, dew point, and temperature. Relative humidity is the amount of water vapor dissolved in the air relative to the amount the air can hold at that temperature, usually given as a percentage. For example, 50 percent relative humidity means that for a given temperature, the amount of water vapor in the air is 50 percent, or half, the amount the air can hold before water will condense out as visible droplets. If you change the air temperature, the relative humidity will also change. Warmer air can hold more moisture, so raising the air temperature will decrease the relative humidity. Conversely, lowering the air temperature will increase the relative humidity. If you lower the air temperature enough, the air will reach the *dew point*, or the (low) temperature at which water vapor reaches saturation in the air. Put another way, at the dew point temperature, relative humidity reaches 100 percent. Dew point, air temperature, and relative humidity are inextricably linked such that if you know two, you can always calculate the third.

vapor barrier (see resources, page 237). From one perspective, the vapor barrier should go on the side that's subject to more vapor drive throughout the year. In many climates, the summer will generate more condensation simply because warm air can hold more moisture than cold air. From another perspective, ventilation should go on the side of the wall that will have an easier time drying. In many situations, this is the outside of

Condensation: In Versus On

It's worth differentiating between condensation that occurs *within* a winter storage room's walls and ceiling and condensation that forms *on* its walls and ceiling. Both can cause problems, but they stem from different issues.

Condensation on the interior wall and ceiling coverings is common in high-humidity storage rooms and occurs anywhere the temperature of wall surface drops below the dew point. If there's inadequate insulation or thermal bridging that hasn't been addressed, you'll likely see this type of condensation. At very high relative humidity, the margin between the air temperature and the dew point is often small. For example, if the air temperature is 32°F (0°C) with a relative humidity of 95 percent, the dew point is 30.7°F (−0.7°C), only 1.3°F, or 0.7°C, away from the air temperature. If you live in a harshly cold climate, you'll likely experience this form of condensation without ample insulation and in places where thermal bridging occurs. During a recent cold snap at my farm in Fairbanks, outdoor temps hovered between −20 and −40°F for two weeks. Inside my storage room, a thin layer of ice crystals formed over some interior walls because the walls were below both the dew point and the freezing point. (The veggies were

fine because of air movement and space between containers and the walls.)

Interior condensation isn't always a problem; it depends on what covers the interior walls. OSB and gypsum drywall, for example, may swell and soften, whereas sheet metal or FRP panels will not be affected. You may also see condensation collecting on metal fasteners set into wall studs and ceiling joists. During the warm season, I've even seen condensation form on metal fasteners in the path of cold air coming from the evaporator; nail heads were cooled below the dew point, and enough condensation formed to cause dripping.

Condensation within walls and ceilings occurs when water vapor gets inside and condenses where the temperature reaches the dew point. Air movement through gaps and cracks is the main conduit for water vapor into walls and ceilings, but water vapor can also diffuse straight through many materials, including drywall and plywood. Some condensation is inevitable unless your vapor barriers are perfect, which is very difficult to achieve in a lasting way. This is why it's important to choose building materials that can handle some level of dampening. It's also important to think about where condensation is likely to form and to give the moisture a path out.

the building. On the ventilated side, you will improve the ability of the wall to dry if you provide a small but continuous air gap between the insulation and the final layer of sheathing (or interior finish material). Perhaps the best advice I can give is to avoid the conundrum entirely and use materials such as MIPs or prefab cooler panels—in other words, materials that provide inherently perfect vapor barriers and don't mind getting wet

at the seams—for any storage spaces that will pull double duty in summer and winter.

While different materials are better or worse at slowing the movement of water vapor, few can stop it completely. That is to say, few materials used in construction are true vapor barriers (despite the common use of the term). Glass and aluminum foil are examples of true vapor barrier materials; water vapor

These wall vents help alleviate some condensation buildup that inevitably happens during the summertime. Even when the storage room isn't actively cooled, it can be cold enough inside to cause humid outside air to condense in the walls. These vents give that moisture a path out. *Photo courtesy of Phil Knapp.*

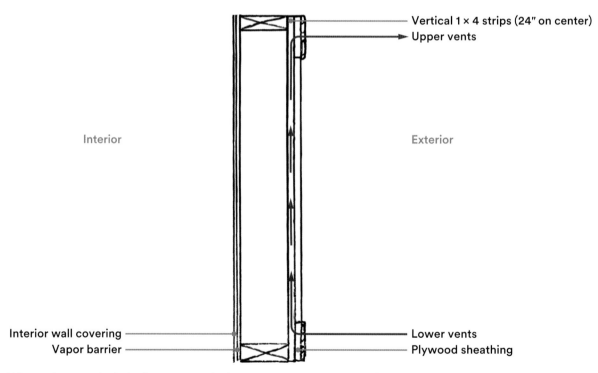

When using a cooler in both summer and winter, one way to manage moisture is to place a vapor barrier on one side and vents plus an air space for convective drying on the other. In this diagram, the vapor barrier sits on the inside of a standard-framed and insulated wall. An air gap created by fastening 1 × 4 strips vertically and spanning the height of the wall allows for convective drying of summertime condensation through vents placed along the bottom and top of the wall.

cannot diffuse through them. Most other materials have varying levels of permeability. Permeance is a measure of the permeability of different materials to water vapor, and *perms* is a common unit. (In the United States, one perm is equal to one grain of water vapor—a grain is one seven-thousandth of a pound—passing through a square foot of material per hour when the difference in partial pressures is one inch of mercury on opposite sides of the material. Archaic, I know.) While the definition of a perm may not be useful, knowing different materials' relative permeance to water vapor is.

Table 7.1 lists some common building materials and their water-vapor permeance. Polyethylene sheeting at 6-mil thickness (0.06 perms) and foil-faced polyiso panels (0.05 perms) are two of the most common vapor barriers used in construction, but some other materials are surprisingly impermeable to water vapor. For example, dry ½-inch CDX plywood (0.5 perms) is less permeable than an inch of XPS foam (1.1 perms). Two-and-a-half inches of closed-cell SPF are relatively tight at 0.8 perms, while 4 inches of mineral wool are comparatively drafty at 29 perms. Tyvek HomeWrap might as well be a sieve at 77 perms. Notice that the thickness of a material affects its permeability and how some materials become more permeable to water vapor when damp. When designing your winter storage spaces, you should try to use materials with less than 1 perm (even when wet) as your intentional vapor barriers.

I can't speak for elsewhere, but in subarctic Alaska, where vapor barriers are important parts of home construction, 6-mil polyethylene sheeting is the industry-standard vapor-barrier material. Ensuring perfect air seals is important, so flexible and durable polyurethane caulking is used—most often Tremco Acoustical Tile Sealant—for sealing poly sheeting to itself or to various construction materials. Construction seaming tape, which is often red in color, seals cuts and other penetration points, such as staples or vent holes. In their guide to walk-in cooler construction, Store It Cold suggests foil-faced foam board

Table 7.1. Water-Vapor Permeance for Common Building Materials

Material	US perms
Aluminum foil, 1 mil (¹⁄₁₀₀₀ inch)	0.00
Polyethylene sheeting, 6 mil	0.06
XPS, 1 inch	1.1
XPS, 2 inches	0.55
EPS, 1 inch	2.0–6.0
Polyiso, 1-inch foil faced	0.05
Polyiso, 1-inch unfaced	26.0
Mineral wool, 4 inches	29.0
Fiberglass insulation, 4 inches	29.0
Closed-cell spray foam, 1 inch	1.9–2.5
Open-cell spray foam, 1 inch	3.2
Softwood lumber, 1 inch	0.4–5.4
CDX plywood, ½ inch	0.5 (Up to 20 when damp)
OSB plywood, ⁷⁄₁₆ inch	2.0 (Up to 12 when damp)
Concrete, 1 inch	3.2
Tyvek HouseWrap	77.0
Gypsum drywall	50.0

Source: Martin Holladay, "All About Vapor Diffusion," Green Building Advisor, June 12, 2015, https://www.greenbuilding advisor.com/article/all-about-vapor-diffusion.

sealed with spray foam and HVAC foil tape at the seams as a vapor barrier (see resources, page 237).

In practice, most vapor retarders also act as air barriers. Air leakage will deliver much more moisture into your walls—not to mention hurt your overall

R-value—than vapor diffusion. Thus, when installing a vapor barrier, you want the seal to be as tight as possible. For framed walls, this means sealing between the base plate and foundation, as well as creating overlapping seals where walls meet ceilings.

Electrical

I don't want to comment too much on electrical systems since I'm not an electrician and there are real safety concerns when it comes to working with electricity. That said, I want to bring a couple things to your attention. First, you'll likely need a disproportionately large number of GFCI (ground-fault circuit interrupter) outlets. These are the type that cut power quickly and stop you from being electrocuted if you, say, toss a hair dryer into a bathtub. Building codes require them anywhere high moisture is possible. Cold-storage rooms themselves can be unusually damp, and relatively cold wiring can build up lots of condensation. There's also potential for water use in adjacent packing and washing rooms. These are all places where GFCI outlets are a good idea and probably required by code. One problem, however, is that GFCI outlets are expensive! They're sometimes ten times the cost of standard receptacles. In some situations, you can save money by using GFCI breakers instead. These allow you to give the same GFCI protection to an entire circuit while using less expensive, standard receptacles. Consider a GFCI breaker for any circuit that will have four or more GFCI outlets.

At some point during construction, you may need to run power to a new or retrofitted structure. If you might ever install commercial-grade refrigeration (in other words, go beyond CoolBot systems), you'll want to consider installing three-phase rather than single-phase power to your building. You'll need three-phase power to run any machinery or equipment that runs on 440 volts, and it allows for more powerful refrigeration units that pull less humidity from the air. Running three-phase power is more expensive, since you'll need to run four or five wires (three hots, one neutral, which is optional, and a

ground) to the building rather than three wires (one hot, one neutral, and a ground). You may also need an upgraded line drop from your electrical company. That said, it may save you money in the long run, because equipment running on three phases generally draws less current. Again, I'm not an electrician; please talk to a real electrician and your electric company to get the right type of power to your building.

Water and Drains

Winter storage buildings don't necessarily need water access, but you'll find water extremely convenient if you have it. If there's water in your storage and packing building, you can use the space for any necessary wintertime washing and also feed water to your humidification equipment. A direct line to a well head is the most convenient option, and necessary if you're frequently running washing equipment, but holding tanks are options when water use is less frequent and lower-volume. A holding tank might also be a good option for people who eventually want to install a well but can't afford one quite yet. Holding tanks are very common in interior Alaska because of the high prevalence of permafrost and arsenic in the ground water. Sometimes they're buried underground or kept in heated sheds, and they can usually hold 1,000 to 2,000 gallons. Smaller tanks might fit into interior spaces within a storage building. For example, I found a 270-gallon stainless-steel tank that was originally used in a coffee shack. Whether you're running water from a well or a tank, you'll likely need to include space for a pressure tank unless you're piping water from an existing pressure tank in another building.

If you're using water inside, you'll also need drains. Rinse water from hand-washing stations or vegetable washers likely won't require a full septic system and can, depending on local codes, be piped away through floor drains into leach pits or onto the soil surface. Floor drains and piping are usually installed before concrete slabs. It's important that slabs pitch slightly toward floor drains so that water doesn't pool elsewhere. If you'll be washing veggies inside or over drains,

The sediment trap system for Food Farm's barrel washer allows sediment to settle out of the water in multiple stages to prevent it from clogging the drains during wintertime washing. *Photo courtesy of Janaki Fisher-Merritt.*

you'll also want to think through methods for sediment trapping and removal. If you don't, sediment might build up and eventually clog your drains, rendering them useless. Some washing equipment includes sediment traps, and there are specific floor-drain products designed to trap sediment. The basic idea is to allow sand and other sediment to settle out before water moves further down the system. Whatever you use, make sure it's easy to access and clean. Certain types of channel drains can trap excess sediment, and P-traps—so long as there's an access port for cleaning—can trap additional sediment as well. It's also wise to install a drain cleanout, which allows you to insert pipe snakes or hoses to remove obstructions further down the system. Keep in mind that if you install bathrooms, you'll need a septic system, which will add thousands of dollars to the overall project cost.

Floorplans and Dimensions

While you can lay out your winter storage spaces however you'd like, you'll want to keep a few things in mind

as you design them. The easiest and least expensive shape to build is a rectangle. Any deviations from or additions to that basic shape will complicate the construction, especially with the foundation and the roof. The size and dimensions of your storage space(s) should at least meet your needs for the crops and amounts you grow. If your packing area is part of the same structure (trust me, you'll appreciate this), you'll need space to maneuver handling equipment, lay out different vegetables and orders you're working on, store supplies and containers not currently in use, and move freely around it all. Intentionally design access pathways into a space. Barn door– and overhead-style doors can help you better utilize available space.

While designing my storage building, I found it helpful to use scale drawings of the floor plan and draw (to scale) objects that might occupy space in the packing room: tables and counters, storage racks and shelves, unused containers, sinks, and so on. This was also helpful for designing the space with workflows in mind. I've read a lot of advice about creating linear workflows through spaces; you don't want to be constantly moving things around as you work, after all. Don't be afraid, however, to use the same space for different tasks at different times of the year. For example, I use a set of racks in the back of my storage building for storing squash in the late fall and early winter, and I use these same racks for seed starting in the spring. The overhead door that's crucial for bringing veggies into storage during harvest doesn't see much use in the winter and instead is the perfect nook for parking my BCS snowblower. While I appreciate when every tool has its place, I acknowledge that spaces built on limited budgets need to be adaptable. Be creative and think about how your investment in a building might serve you at all times of the year.

You may need multiple storage spaces to provide different temperatures and humidities for different crops. Refer back to the different storage groups to see what conditions each group needs (see "Storage Room Conditions," page 45). There can be overlap between some of these groups, and you can sometimes get away with storing certain crops in less-than-optimal conditions. It will shorten the storage lives of those crops, but many farmers, myself included, accept this fact. For example, I keep my onions and garlic in the space conditioned for my winter squash (see "Offbeet Farm," page 183). Because I don't grow large amounts of onions and garlic, I'm able to sell them within three to four months of harvest before I see issues with sprouting. West Farm in Vermont stores winter squash and sweet potatoes together to save space (see "West Farm," page 203). As general rules, try to avoid storing easily chilled crops at temperatures that injure them; store alliums in dry locations, preferably under 70 percent relative humidity; don't store members of the cold and damp group above 40°F (4°C); and in dry storage rooms, provide semi-enclosed containers to keep moisture-loving crops from drying out.

Storage Room Minimum Dimensions
How big does each storage space need to be? Each space will need to fit all the storage containers it takes to hold veggies after harvest (see "Storage Container Needs," page 51, and "Arranging Storage Containers," page 52). In addition to the total number of storage containers, you'll need to know how those containers fit together in space. This means answering questions like how high to stack containers (if applicable) and how to shape the storage rooms to be most functional.

How high can you comfortably stack your containers? Some bins will have a maximum weight capacity when stacked—that is, the maximum load the container on the bottom can handle safely. Be sure to look for this information when purchasing containers. If you're building containers yourself, design them to be strong when stacked. You might also be limited by the capacity or maximum lift height of a pallet stacker. Make sure your stacks are stable! Unstable stacks are not only dangerous to you, your crew, and your facility; a fallen stack could potentially damage thousands of dollars of veggies and cause time-consuming cleanups. Once you've decided on the number of containers in each stack, you can figure out the total height. Be sure to include the height of any blocks or pallets elevating containers off the ground.

The number of stacks will help determine the minimum dimensions for your storage rooms. How many stacks of containers will you need to hold the harvest? Divide the total number of storage containers needed by the number in each stack. Returning to the example of carrots in 27-gallon totes that we used to calculate container needs in chapter 4 (see "Storage Container Needs," page 51), if you stack the bins five-high, you'll need eleven stacks—ten full stacks and one stack of three totes—to accommodate the fifty-three totes needed to hold 5,000 pounds of carrots. Next, think about the ways those stacks can be arranged. You could have two rows of six or three rows of four to fit the eleven stacks of containers (with one space empty). The arrangement you choose might be affected by the types of containers you're using, the number of crops you're storing, or limitations of the building site. If you're planning to lift and move your containers with a pallet jack, stacker, forklift, or tractor with forks, it's important to know whether pallets can be lifted from one or both sides, as this will affect the orientation of your containers relative to the door and access aisle(s). Whatever configuration you choose, you now know the necessary dimensions in terms of storage containers.

You can convert this information into useful measurements by multiplying the measure of the storage containers—the width, length, and height—by the number of planned containers in each corresponding dimension. If there will be partial rows or stacks of containers, pretend they're fully filled for these calculations. These dimensions represent the bare minimum length, width, and height of your storage room if no extra space is left for access or airflow.

Airflow, Access, and Future Growth

Airflow in storage rooms is important, however! Without space for air movement, you'll have a difficult time removing field heat at harvest time and preventing cold spots in the depths of winter. There are several general rules when designing extra space for air movement. First, storage containers or bulk piles should not block air coming from the evaporator of your cooling system (including CoolBot systems). If containers or piled veggies are too high, they may disrupt airflow within the room. Additionally, produce might inadvertently freeze if left in the direct path of air coming from your cooling system. At a bare minimum, there should be about 12 inches (30 cm) between the tops of containers and the ceiling.[4] There should also be space between the containers and walls and between the containers themselves: 8 to 10 inches (20 to 25 cm) on endwalls, 4 to 6 inches (10 to 15 cm) on sidewalls, and 4 inches (10 cm) between pallet-sized stacks of containers.[5] These spaces facilitate cooling and protect against freezing near walls. In cold-storage parlance, endwalls are those in the direct path of airflow, facing directly toward or away from evaporator and/or ventilation fans, whereas sidewalls are those set parallel to the path of airflow. Endwalls usually face the longer dimension of a room, while sidewalls usually face the shorter dimension. It's also a good idea to keep containers elevated off the floor with pallets or planks if your

containers don't include an airspace underneath as part of their design. Standard wooden pallets are 5.5 inches thick, and it's good practice to orient them with the openings in the direction of airflow. Good airflow beneath containers will help prevent freezing near the floor. By raising containers off the floor, you might also avoid damage should the storage space ever flood.

It's also important to think about how you will access the storage room to move things in and out. If you're storing two or more crops in the same room, you should be able to access all crops easily without moving containers around. One way to achieve this is to build a long, narrow storage room with an access aisle running along one wall. Another is to design a square or rectangular room with access aisles running along two walls. A third option that maximizes usable floor space is to place an access aisle centrally in a storage room to grant access to two rows of containers at once. In all these scenarios, the location and size of the door will affect where it makes sense to place access aisles.

If your storage containers are small enough to carry by hand or with a dolly, you'll need to leave at least a

Food Farm maintains a central aisle in their cold-storage room to access bins on two sides, but sometimes that space is temporarily needed for packed orders before they go out the door. *Photo courtesy of Janaki Fisher-Merritt.*

container's width of extra space in the aisle, and preferably slightly more, to avoid constantly bumping into walls or other containers. If the storage containers are large enough to need pallet jacks, stackers, or forklifts, you need enough space to account for both the containers and the handling equipment. Determine which is greater: the diagonal corner-to-corner width of your storage containers or the front-to-back length (from the tips of the forks/platform to the back of the tool) of your handling equipment. For the latter measurement, include any additional space required to operate the lifting equipment. Making your access aisle(s) at least as wide as the greater of these two lengths will ensure two things: (1) that you can turn a container into an empty spot anywhere along the aisle and (2) that you can turn and face a row with the container(s) in the fully lowered position. This second part is important for safely moving containers and creating stacks. If your handling equipment is very large and requires a lot of space—a tractor with forks, for example—consider including large access doors oriented so that equipment can drive straight in and set down containers without turning. Depending on the size of the access door, you might be able to avoid large amounts of unused access space if the handling equipment can lift containers while partially sticking out the door.

You can add the additional spaces needed for both access and airflow to the room dimensions calculated in the previous section. The results are the minimum interior dimensions of a storage room that will allow you to meet your sales goals for long-term winter storage. That said, there are many, often creative ways to arrange containers and use floor space to maximize usage and efficiency in a cold-storage room. Consider rotating certain containers to fit better in a given spot or using multiple container sizes to get the most out of tight spaces. Think about the location and size of the door to the storage room and how its placement might change possible arrangements of containers. Think about the order in which crops and containers might enter and leave the storage

room. By thinking creatively about your unique situation, you may be able to reduce the necessary size of your storage room and lower the upfront infrastructure costs for the farm.

Is the "smallest storage room possible" really what you want, though? In many cases, no. Harvest yields vary from year to year, so it's prudent to include some extra space to accommodate a bumper crop. You may also want to account for the farm's potential growth. Will the amount of winter storage and sales change in the next three, five, or ten years? If so, think about your potential to expand and consider building to that larger capacity right away. Extra winter storage space can also be a source of revenue if you rent that unused space to other farms. That said, the size and use of your storage space might depend on your funding sources. Certain FSA loans, for example, won't allow you to build extra capacity or store produce from other farms.

Building Access

Wherever you decide to build or place your winter storage space, you should consider accessibility relative to the fields, driveways, and roads, as well as how snow, ice, and rain-softened ground will affect access. This might mean pouring additional gravel to shore up soft ground or factoring in snow removal paths. Preferably, your storage space will be close to where you wash veggies, regardless of when you wash them. You will spend a lot of time moving crops between these spots, and small time savings will add up over days, weeks, and years. When thinking about access, consider the types of equipment or vehicles you might use for storage crops. What will the driveway look like? How will you turn a vehicle around? The turning radius for even small pickup trucks will be 20 to 25 feet (about 6 to 8 m). If you don't have enough space for a large pad, pull-outs for making three-point turns can be helpful for getting vehicles turned around. Will your building have a loading dock? If so, there must be space for your delivery vehicle to quickly and easily back up to the dock.

To Hire a Contractor or Not

As farmers, we often feel compelled to complete building or repair projects ourselves without hiring outside help. This can be necessary, especially if the farm is relatively new; without years of built-up capital or collateral for loans, we can't always afford to hire professionals. That said, building a vegetable storage space is a multifaceted project that might push us beyond our current skill sets. That leaves us with a decision: to hire a contractor or not when building a winter storage facility.

Before tackling this question, it's important to understand the role of a contractor in a building project and what services they'll provide. Contractors organize and coordinate most aspects of a project. In the context of music, they'd be the composers, conductors, and event organizers all in one. Some might even play in the band. Some contractors take part in the design process and either create blueprints or coordinate with an architect to make certified building plans. (If not, you might have to create designs with an architect first before approaching a contractor.) Contractors are responsible for ordering all the materials for the project and getting them to the building site. Many contractors do much of the work themselves or with a core crew they hire. This is especially true for small, independent contractors who run their own business. Contractors hire other, more specialized businesses—often called subcontractors—to perform work outside the core crew's purview. This is often the

Friends of Offbeet Farm help add cross-bracing to the roof trusses.

case for concrete, plumbing, and electrical work. If you hire a licensed, bonded, and insured contractor—and I wouldn't hire one who isn't—they accept a certain amount of responsibility for the project and ensure it meets local codes. If, for example, someone gets injured during construction, the contractor's insurance will pay for the medical costs. If some roof trusses fall and break during installation, the contractor will pay for replacing the broken pieces.

While these are great services, they come at a price. Hiring a contractor can more than double the project cost. Aside from the cost of their time—many contractors have steep hourly rates for boots-on-the-ground and planning time—and the services of the subcontractors, some contractors tack on percentage fees for all materials purchased during the project. Most, however, will make these costs clear upfront and present you with itemized budgets for the proposed projects before you start. If you have a limited budget, you need to make this clear. Contractors generally aren't constrained by their initial estimates and might blow through your budget if left unchecked.

In contrast, tackling a project yourself means that you're in charge. It's your responsibility to order and transport materials, obtain local building permits and meet codes, and hire and manage subcontractors for any work you can't do yourself. If you're getting a loan, you'll have to generate materials budgets and get bids from subcontractors to present to the bank. You'll also need to convince your lender that you're capable of completing the project yourself. Your loan officer will have the final say, and you might have to provide evidence of past work. If someone gets injured during construction, either you or your insurance will be responsible for the medical costs, and if there's a defect in the construction or the work doesn't meet local codes, there's no one to blame but yourself. Construction also takes a lot of time, which means time away from farming. Unfortunately, prime construction season overlaps with the growing season in many locations. That said, it's probably cheaper to hire an extra farmhand or two than to hire a contractor to build something for you.

The final decision depends on many things, but I think it boils down to time, money, your skills, and your willingness to learn new things. If you have the skills and the time, or if you're willing to learn new skills and devote the time to doing it, you may want to try building something yourself. This is your only option, after all, if you can't afford a contractor. Some contractors may be open to a middle ground in which you perform some aspects of the project yourself or work alongside them. This will lower the total cost while putting the overall responsibility in more capable hands. You might be surprised with what you can accomplish yourself, though.

Chapter 8

Climate Control

The question I'm asked most as a storage farmer: "It's January (or another winter month)! How do your veggies look so good?" The answer is that I provide the right conditions in storage. Climate control is where the rubber meets the road; providing the right temperature and humidity is paramount to maintaining the quality and longevity of your vegetables in long-term winter storage. But how do you achieve that? If you have a well-designed and well-built storage space, effective climate control is about having the right systems and equipment in place to keep conditions where you need them. Your goals are to keep crops dormant while retaining moisture, preventing pathogen growth, and preventing injury from cold. As I've mentioned before, there can be a little wiggle room around the right temperatures and humidities for certain crops, but nailing the ideal conditions will give you the greatest chances of success.

Controlling Temperature

We need to control the temperature in our storage spaces because ambient conditions of outside air and soil are rarely conducive to veggie storage. Most often, your storage spaces will need refrigeration, but you might also need heating depending on your climate, your facilities, and your crops.

Table 8.1. Ballpark Cooling Loads for a Winter Storage Cooler at 32°F to 35°F

Room Dimensions	Volume (ft³)	Estimated Cooling Load (BTUh)
8′ × 10′ × 8′	640	9,000
12′ × 20′ × 8′	1,920	18,000
16′ × 24′ × 8′	3,072	25,000
24′ × 30′ × 8′	5,760	40,000

Note: These estimates are approximate averages from several sources and assume conditions common to storage farms at peak harvest: good levels of insulation, hot weather, heavy cooler use, and newly harvested veggies entering with field heat. For 10-foot ceilings, add 15 percent to the cooling loads.

Sources: "Quick Calculations for Walk-in Coolers and Freezers," Heatcraft Refrigeration Products LLC, https://www.uscooler.com/blog/load-calc.pdf; "Load Calculator," KeepRite Refrigeration, https://k-rp.com/load-calculator; "Refrigeration Sizing Estimate," U.S. Cooler, https://www.uscooler.com/refrigeration-sizing-estimate; Scott A. Sanford and John Hendrickson, *On-Farm Cold Storage of Fall-Harvested Fruit and Vegetable Crops: Planning, Design and Operation* (University of Wisconsin-Madison, Cooperative Extension, 2015); "Quick Calculations for Walk-in Coolers and Freezers," Heatcraft Refrigeration Products LLC, https://www.uscooler.com/blog/load-calc.pdf.

Your refrigeration systems need the capacity to remove field heat from potentially thousands of pounds of vegetables at a time. *Photo courtesy of Amy Frye.*

Cooling Loads

Your cooling equipment needs to work against many sources of heat to maintain proper temperatures in storage. The largest source of heat will be the outside environment, which includes not only air but the soil under (and sometimes around) your storage space, as well as direct sunlight, which increases the effective outdoor temperature around your cooler. Your storage rooms can also absorb heat from adjacent indoor spaces through walls, floors, and air entering through open doors. During harvest, you'll need to remove field heat carried into the storage room by freshly harvested veggies. Even after veggies come down to storage temps, your cooling system must remove the constant stream of heat they release via respiration. As if that's not enough, lights in the storage room release heat, and so do any people inside. Even the motors running the refrigeration system release heat into the storage space.

So how much cooling power do you need? As you've probably guessed, it's complicated. Not only do you have to consider all the things above, but there's also the amounts and location of insulation, changing temperatures outside, harvest rates, how often you're opening the doors, and the daily runtime of your cooling system, including defrost cycles. There are whole books written about calculating refrigeration loads for coolers. I highly recommend either hiring a refrigeration specialist to do the load calculations or learning to do them yourself. There are great online resources for learning simplified cooling load calculations, including information prepared by the National Cooperative Extension system (see resources, page 237). To get you started, Table 8.1 gives some ballpark estimates for cooling loads on storage farms during harvest. And you'll want to design cooling systems with the worst-case scenario in mind—that is, the hottest weather during harvest with the highest possible harvest rate. It's better to have an oversized system than an undersized one.

The Basics of Cooling Equipment

Regardless of size or cost, most cooling systems work with the same basic principles and components. They typically consist of two sets of connected coils filled with refrigerant. At one coil, the refrigerant absorbs heat from the space you're cooling; this is the evaporator. A fan pulls air from within the storage room through the evaporator coils to blow colder air back into the room. At the other coil, the refrigerant releases the heat absorbed at the evaporator; this is the condenser. There, a fan blows ambient air over the condenser coils to remove the collected heat. Both the evaporator and condenser coils are surrounded by fine metal fins to

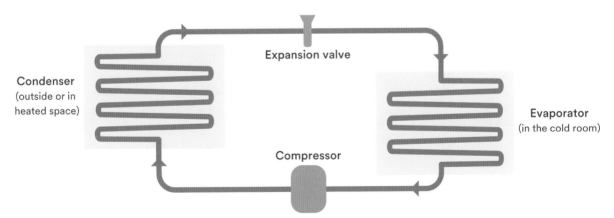

Most refrigeration systems have four main components for transferring heat from one place to another: an evaporator to remove heat from a space, a compressor to pressurize gaseous refrigerant, a condenser to dump heat into another space and condense the refrigerant into a liquid, and an expansion valve to turn refrigerant from a liquid into a gas.

assist with heat exchange. Two more components make this heat transfer possible. One is the compressor. The refrigerant is in a gaseous state after passing through the evaporator coils, and the compressor raises both the pressure and temperature of the gas. The pressurized refrigerant condenses back to a liquid when it passes through the condenser coils before continuing through an expansion valve. The expansion valve limits flow, which drops the pressure of the refrigerant. The sudden pressure drop forces much of the refrigerant back into a gaseous state, which also lowers its temperature. So long as the refrigerant is colder than cooler-room air at the evaporator and warmer than ambient air at the condenser, the system effectively pulls heat from within the storage room to deposit it elsewhere.

Another thing common to all cooling systems is their effect on humidity inside the storage room: All cooling systems limit humidity to some extent. When air from inside the storage room passes over the cold evaporator coils, water vapor in the air condenses on the coils if they are at or below the dew point. This condensation also freezes if the coils are below 32°F.

Table 8.2. Relative Humidity Limits for Coolers at Different Split Temperatures

Temperature Drop Across Evaporator (°F)	RH Limit at 32°F Room Temp (%)	RH Limit at 38°F Room Temp (%)
–1°	96	96
–2°	91	92
–3°	87	89
–4°	83	85
–5°	79	82
–10°	63	65
–15°	49	49

Source: James A. Bartsch and David G. Blandpied, *Refrigeration and Controlled Atmosphere Storage for Horticultural Crops* (Ithaca, NY: Northeast Regional Agricultural Engineering Service, 1990).

The temperature difference between air in the cooler and air coming off the evaporators is sometimes called the split temperature or the temperature drop across the evaporator. (Note, this does not take the temperature of refrigerant in the coils into account; it's just air temperatures. To measure the split temperature, measure the temperatures of air entering and exiting the evaporator and subtract the difference.) If you know the split temperature, you can know how much your cooling system will limit humidity in the room. Basically, the larger the split temperature, the more your cooling system will limit the relative humidity in your cooler. Table 8.2 shows these limitations for different split temperatures in coolers set to 32°F (0°C) and 38°F (3.3°C). Systems with low split temperatures tend to have large evaporators with lots of surface area. These huge evaporators are preferable for vegetable storage, since you will have an easier fight against the dehumidifying effects of your cooling system. There is also less danger that air coming from the evaporator will freeze veggies in its direct path. In the context of vegetable storage, the ideal split temperature is about 1°F (0.6°C), which can maintain relative humidity up to 96 percent (without extra humidification). With larger split temps, you'll need to actively humidify the space to achieve relative humidities above the limitations. That not only means more electricity to run humidifiers; it also lowers the effective cooling power of your system (since heat energy is released when water condenses from a vapor to a liquid on the evaporator coils).

Commercial refrigeration systems come in two flavors: remote and single-unit. With remote systems, the evaporators and condensers are separated. The evaporator coils will be in the cold-storage room, while the condensers will be somewhere else—sometimes outdoors, but sometimes inside, where the waste heat warms an indoor space. Single-unit systems are as they sound: The evaporator and condenser are housed within a single casing that spans either the wall or ceiling of the cooler. You can also avoid commercial refrigeration entirely and use residential air conditioners (ACs) paired with CoolBot controllers. Regardless of type, most cooling systems will require electricity delivered at 208/230 volts and will draw less amperage if using three-phase power.

Before we discuss the pros and cons of the different systems, let's go over the challenges of comparing systems and their cooling capacities. Most self-contained systems and residential ACs will have a rated cooling power in BTUs per hour (BTUh). Unfortunately, you can't compare these numbers apples to apples without first knowing the conditions for which they're rated. Manufacturers rate the cooling capacity of systems for specific indoor (cooler) temperatures and ambient (outdoor) temperatures within their expected operating range. Self-contained commercial systems are usually rated for conditions specific to walk-in coolers—for example, 35°F (2°C) inside and 95°F (35°C) outside. You'll find this information in the specifications for each system. Residential ACs are usually rated around 70°F (21°C) inside and 95°F outside, but it's not always easy to find the exact specs. The reason you can't compare cooling power willy-nilly is that it changes based upon the operating conditions. Generally speaking, as indoor temperature decreases, the cooling power also decreases. Likewise, as outdoor temperature increases, cooling power decreases. Thus, you can't rely on that 24,000 BTUh air conditioner for 24,000 BTUh of power when cooling your storage space to 34°F (1°C)—more on that soon. Unfortunately, remote cooling systems sometimes use entirely different power ratings based on the cooling capacity of the evaporator—measured in tons—and the mechanical power of the compressor—measured in horsepower. The ton is a unit leftover from the days before mechanical refrigeration, when people used ice for their cooling needs. One ton is the amount of heat required to melt 2,000 pounds of ice in 24 hours, which happens to be 12,000 BTUh. There is no specific conversion between the horsepower of the compressor and cooling power in BTUh, but for medium temperature applications—in other words, typical cooler temps—8,000 BTUh per one horsepower is a ballpark conversion. The

ratings for the evaporator and compressor of a remote system need to be correctly matched in order for them to work properly.

Choosing Cooling Equipment

The choice between the different cooling options will depend on your needs for cooling power, your budget, and your preferences for energy usage. Commercial remote systems are the most common on larger storage farms, but I've seen them installed in small storage rooms too. These are your most powerful options and can sometimes handle huge cooling loads. Again, it all depends on the capacity of the evaporators and the mechanical power of the compressors. While the equipment for commercial remote systems can be relatively inexpensive, the installation and potential maintenance costs are not. Most remote systems must be charged with refrigerant and pumped down onsite, which requires licensed refrigeration installers (because of the safety and environmental hazards of working with refrigerants).[1] For context, a local HVAC company recently quoted me $40,000 for a remote system in the range of 15,000 to 20,000 BTUh. Most of that cost was for labor. Some companies sell preassembled remote systems that come precharged with refrigerant. While this will limit your installation options, you'll skip much of the installation cost. Unless you have the technical knowledge to program the system, it's still a good idea to hire an installer to help with the initial setup. If either of these systems ever needs repair, you'll need to call out a licensed technician if the refrigerant must be recovered as part of the job.

Despite their expense, there are some real benefits to installing commercial remote systems. First, they can meet your custom needs for power, temperature, and humidity. Only remote systems have evaporators large enough to maintain very small split temperatures, which will help minimize dehumidifying effects and reduce the risk of frozen produce. Second, you can install the condensers anywhere you want—outside, for example, where the excess heat and noise won't bother you, or inside, where you can capture the waste heat to warm another indoor space.

At the smaller end of commercial refrigeration systems are the self-contained units. Because of practical space limitations for fitting all the components inside a single housing, these units tend to have less power than remote systems. That said, I've seen a model rated up to 36,000 BTUh at 35°F (2°C) indoor and 90°F (32°C) outdoor. (This one weighed over 300 pounds, or 136 kilograms!) Self-contained units come in several designs depending on where you want the evaporator and your needs for space/shelving. Top-mounted units sit on top of the cooler, and the

evaporator drops through the ceiling. These are most commonly used with prefab coolers, and there are systems rated for both indoor and outdoor use. Side-mounted units install through a wall of the cooler, similar to residential ACs. You're more likely to see these used with DIY coolers built into the structure of a building. Lastly, there are saddle units that straddle the top edge of a wall with hanging brackets.

Just like the preassembled remote systems, self-contained units come precharged with refrigerant, so you may not need to hire an installer. They're also rated at their working temperatures, so there's no guesswork about the cooling power. The evaporators

Large evaporators of remote systems can have the surface areas necessary to limit dehumidifying effects on a storage room. *Photo courtesy of Holly Simpson Baldwin.*

Comparing CoolBot and Commercial Refrigeration Systems

CoolBots offer great opportunities for farmers to lower startup costs for commercial cold storage. If you check out the Store It Cold website, you will read how the CoolBot not only lowers purchase and installation costs for refrigeration, but it also lowers energy usage up to 40 percent compared to commercial systems. This can be true, but there's some context worth knowing. I reached out to Store It Cold for more information and was grateful for the chance to interview their support and operations manager, who helped me dig into the specifics of CoolBot systems. Here are some important points I gleaned from our conversation and from my subsequent research.

CoolBot systems have a limited range of operating temperatures. Theoretically, a conventional air conditioner can bring a room down to 32°F. While some customers have reported success in doing so, the company doesn't recommend setting the temperature lower than 34°F (1.1°C). CoolBot systems rely on (relatively) warm air from the storage room to melt ice accumulating on the evaporator coils and fins during defrost cycles. The closer the room gets to the freezing point, the more difficult this becomes, and freeze-ups—when the evaporator coils and fins get encased in ice—become more likely. (Some commercial units have heated defrost cycles to get around this limitation.) If you're only relying upon your refrigeration system for a short time, maintaining a higher-than-optimal temperature might not be a big deal. For example, I only run my CoolBot system for one to two months during and after harvest. After that, I use cold outdoor air to keep my storage room at 32°F. If your climate doesn't allow for this,

you'll have to decide if a storage temperature of 34°F rather than 32°F is acceptable.

I can offer some guidance since my original storage room back at Root Cellar Farm hovered close to 34°F for most of the winter. Typically, I saw higher losses, sprouting, and surface blemishes on my cold-loving crops after four to five months in storage. This compares to my experience at Offbeet Farm—where storage temps stay at 32°F most of the winter—where those same crops stay in good condition an additional one to three months before showing the same issues. And in case you're wondering, it's perfectly fine to run CoolBot systems during the winter, but their efficiency suffers when outside temps drop below 25°F (−4°C). Thankfully, that's a temperature at which cooling with outdoor air is a viable option.

Air conditioners do not operate at their advertised cooling power when working with CoolBot controllers. The reduction in power depends largely on temperature in the storage room. It can be difficult to select the right size and number of ACs to meet your cooling load. That's because, as I mentioned earlier, ACs are rated for indoor temps around 70°F (21°C), and the actual cooling power drops as indoor temperatures decrease. Store It Cold offers some helpful sizing guides on their website (see resources, page 237), but coolers on many storage farms are larger than those on the chart. So, is there a way to compare the power of CoolBot and commercial systems apples to apples?

Kind of. Store It Cold said a general rule of thumb is to divide the advertised cooling power of an AC in half when operating a CoolBot system in the range of 38 to 49°F (3 to 9°C). Thus, with a set temperature of 39°F, a 24,000 BTUh air conditioner should have around 12,000 BTUh of actual power. What about operating at a set temperature of 34°F? The answer is a little murkier. Store It Cold doesn't have any data for this, but the data from a study by the Practical Farmers of Iowa suggest that actual cooling power for ACs might be as low as 10 to 20 percent of what's advertised.[2] From observations on my farms and from comparing cooling load calculations to Store It Cold's charts, I think actual AC cooling power at 34°F is closer to one-fourth or one-third of what's advertised.

There are few limits to the number of Cool-Bot systems operating in a single room. Given the weight of storage crops and the need to store everything at once, most storage farmers have cooling loads large enough to require multiple ACs with CoolBots. I asked Store It Cold if there's a limit to the number of CoolBot-controlled ACs working in a single storage room. The answer: effectively, no. They mentioned some interesting examples of warehouses in central America employing small armies of CoolBots. However, there are pros and cons to deploying a lot of ACs together. On one hand, the more CoolBot systems there are, the more individual components there are to maintain, fix, and replace when broken. On the other hand, with more systems in place, the loss of any one CoolBot-AC

pair becomes less important to maintaining cool temperatures. Regardless of the cooling load, the purchase and installation price for CoolBot systems versus commercial refrigeration will almost always be cheaper when meeting the same load.

Commercial systems can cool a room faster than CoolBot systems. Store It Cold cautioned that commercial refrigeration systems often outperform CoolBot systems when it comes to recovery—in other words, the time it takes to recool a room after warm products or warm air enter. There are multiple reasons for this. One is that commercial systems are beefier—with larger coils, larger fins, and large (and sometimes more) fans—compared to residential ACs. Commercial systems also have built-in defrost cycles, enabling them to spend more time engaging the compressor and less time defrosting. CoolBot systems, by contrast, aren't designed to generate lots of power quickly, and they need to spend more time defrosting, culminating in longer recovery times. This is an argument for oversizing CoolBot systems on storage farms, because quality and longevity in long-term storage depends on quick removal of field heat.

Single-AC CoolBot systems often use less energy than comparable commercial refrigeration. Something Store It Cold prides itself on is reducing energy usage for its customers.

on self-contained units are usually smaller than on remote systems, so there will likely be larger split temperatures. That said, these systems are considerably cheaper to purchase and install, often less than half the cost of comparable remote systems. It depends on the unit, but self-contained commercial

refrigeration systems today sell for anywhere in the range of $4,000 to $10,000.

CoolBot controllers allow people to use residential ACs as refrigeration equipment. The system was developed by New York organic farmer Ron Khosla, who patented the idea in 2006 to start the company

However, the amount of energy savings from CoolBot systems will depend on the context. The study that Store It Cold commonly cites was published by the New York State Energy Research and Development Authority (NYSERDA) in 2009.[3] The energy savings of up to 40 percent came from a scenario comparing a CoolBot system cycling the AC fans on and off to a remote system continuously running its evaporator fans. I should note that continuously running fans helps with system recovery, and Store It Cold recommends that users run their AC fans continuously. Therefore, this scenario isn't a fair comparison. When both systems cycled their fans on and off, the CoolBot system actually used more energy (roughly 14 to 15 percent more) than the conventional system.

Strangely, there's no comparison for the scenario in which both systems run fans continuously, which is the most likely setup. Another study from Iowa compared similar coolers on different farms and found that the CoolBot system used less energy than the conventional refrigeration system.[4] That said, a direct comparison isn't entirely fair because the systems are on different farms. Store It Cold makes a good point on its website, stating that ACs usually run a single 300-watt evaporator fan, whereas many commercial systems run multiple evaporator fans. I should note, though, that not all commercial refrigeration runs multiple evaporator fans—self-contained systems, for example, do not—so again, comparative energy usage will depend on the context.

It's unclear how energy usage with multiple-AC CoolBot systems stacks up against comparable commercial systems. Unfortunately, Store It Cold doesn't offer any data comparing energy usage with conventional refrigeration versus CoolBot systems using multiple ACs. Likewise, I couldn't find any published studies making this direct comparison. Given that many small-scale farms need multiple-AC CoolBot systems, I'm going to shamelessly plug that some research organization conduct this study! I gleaned some information from the Practical Farmers of Iowa study, which profiled a farm using four CoolBot-controlled ACs in their cooler. From that data, it's clear that energy usage with multiple-AC CoolBot systems isn't purely additive. In other words, four AC-CoolBot combos don't use four times the energy of one AC and CoolBot. It's less—likely because of reduced cycling with multiple air conditioners working together on the same space. It's hard to say how that energy usage compares to a comparably powered commercial system. I ran some calculations and found similar amounts of energy usage for the two systems, but I had to make too many assumptions to fully trust the results.

Store It Cold. The CoolBot has been a boon for not only farmers but for people needing small-scale refrigeration across the world. It has effectively lowered the bar for entry into small-scale commercial refrigeration. An 8,000 to 10,000 BTUh commercial system that costs around $4,000 might be replaced by an air conditioner and CoolBot controller for less than $1,500. You're probably intrigued, but there are some important limitations and factors to consider when comparing CoolBot systems to conventional refrigeration (see "Comparing CoolBot and Commercial Refrigeration Systems," page 113). While

Storage farms often have large enough cooling loads to require multiple-AC CoolBot systems. *Photo courtesy of Allison Stawara.*

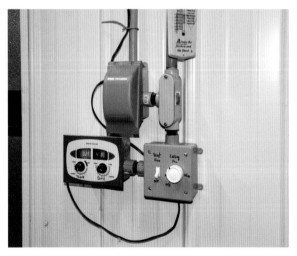

Differential thermostats are one method for controlling intake fans to cool spaces with outdoor air. By adjusting the differential, you have some control over humidity in the room.

CoolBots are often the least expensive option, there may be practical reasons—space, maintenance, and temperature requirements, to name a few—to choose more conventional refrigeration instead.

Cooling with Outdoor Air

Many storage farmers find that their cold-storage rooms need refrigeration even when outdoor temperatures are well below freezing. In such circumstances, or anytime that outdoor temperatures are below the desired storage temp, outside air can be a smart alternative for refrigeration. This will save energy because you only need fans rather than full cooling systems with compressors and evaporators. The setup is relatively simple, consisting of one or multiple intake fans controlled by thermostats. You should also install a passive air vent (or a second fan blowing out) to help exchange warm and cool air and prevent positive air pressure in the storage space. The fans should be rated to handle humid air, like bathroom vents or duct fans for greenhouses. Thermostats will turn the fan(s) on and off based on both inside and outside temperatures to hold the room within a chosen range. One way to do this is with a differential thermostat like those used in solar water heating applications (see "Food Farm," page 219). The differential thermostat will monitor both indoor and outdoor temperature, plus the difference between them (the differential), and allow you to program set points for both the indoor and differential temperatures. When the inside temperature and the indoor-outdoor differential are both above the set points, the fan turns on until the inside temperature drops back to the set point. The differential allows for some control of humidity in the storage room, since outdoor air is usually drier than air in the storage room.

Another setup involves connecting two thermostatically controlled outlets (such as InkBirds) in series. Here, series means the controllers and fans are connected end to end (in other words, in series): You plug (or wire) one controller into a receptacle, plug (or wire) the second controller into the first, and plug (or wire) the fan(s) into the second controller. In this

116

The intake fan (connected to 6-inch flexible duct) is regulated by two thermostat controllers connected in series. This setup saves energy in the shoulder seasons by taking pressure off the CoolBot system.

Digital Temperature Controllers

There are many types of thermostats potentially useful for winter vegetable storage. A particularly useful type are digital temperature controllers. Many of these allow users to plug an appliance into an integrated outlet and set a desired temperature range for that appliance to operate. InkBird is a popular brand for temperature controllers among small-scale farmers because of their low price, ease of use, and reliability. On my farm, I use InkBird controllers to run my air-intake fans and my supplemental heater. You can easily calibrate them to 32°F with a container of ice water, and they can handle outputs up to 2 kW.

setup, both controllers need to be active/on for the fan to run. Set the thermostat monitoring indoor temperature to cooling mode, and program the upper and lower limits for the storage-room temperature. I usually set the upper limit at 33°F (0.6°C) and the lower limit at 32°F (0°C) in my storage room. Set the thermostat monitoring outdoor temperature to heating mode. The upper limit should be the highest air temperature you want to bring into the storage room. In practice, I use the same temperature as the upper bound of the cooling-mode (indoor) thermostat. For example, if I don't want my storage room above 33°F, I set the upper bound of the heating-mode (outdoor) thermostat to 33°F, too. The lower bound temperature for the heating-mode (outdoor) thermostat can be near the threshold for needing supplemental heat. I usually set the lower bound on mine to −10°F (−23.3°C), which ensures the system turns on only when I need it. I've discussed this setup with an electrical engineer, who assured me there's a way to program a single controller—such as a

Raspberry Pi—to do this job, but it's beyond my current knowledge and skillset.

Why not just use one thermostat with the fan? You could do this, just monitoring the inside temperature, but you'd lose some control around the shoulder seasons. I find the two-thermostat setup particularly helpful when outdoor air temps are in the twenties and thirties (about −7°C to 4°C); I can supplement my CoolBot system to save energy without accidently pulling in air that's too warm. Any colder outside and I can cool my storage room with outdoor air alone.

Heating

Many storage farmers engage their refrigeration systems for most of the winter to keep their cold-storage rooms sufficiently cool. However, there are occasions when it's necessary to add heat. This is especially true in very cold climates and for poorly insulated storage spaces. In most cases, it doesn't take much added heat to keep the room above freezing during a cold snap. I use a 1,500-watt space heater plugged into an InkBird thermostat. Several of the farms profiled in part 4 of this book do something similar. In my storage room, I find that supplemental heat isn't necessary until outdoor temps drop to between −10 and −20°F (about −23 and −29°C).

For heating warmer spaces, such as those used for processing orders or storing winter squash and sweet potatoes, you might need a system with more power. Options abound, from more-powerful electric heaters to propane burners to heating-oil furnaces. I caution against using radiant heaters in spaces that store veggies, though. Veggies closer to the radiant heat source will get warmer than those further away (because radiant heat transfer decreases with distance), which can lead to uneven spoilage or sprouting in storage. Whatever you choose, make sure the system is reliable and safe. It's also important to have a backup in place in case you lose power! Although it's not a viable long-term option, I have a small generator to power a space heater and thermostat to keep everything above freezing for a short time should the need arise.

There are also some neat imaginative options for heating. Heat pumps have me more and more intrigued. The technology is only improving, and there are now air-source heat pump systems rated to work even at sub-zero temperatures (Fahrenheit). You can also design your cooling systems to spill waste heat into other spaces. (Refrigeration systems are essentially heat pumps, after all.) So long as you don't mind the noise, putting condenser coils inside is a great way to get "free" heat from refrigeration that you're using anyway. I've even seen more-intentional designs where air-to-air or air-to-water heat pumps cool one space while heating another.

Humidity Control

Alongside temperature, humidity can make or break your results in storage. Too little humidity, and you'll see veggies losing water weight, desiccating, going soft, and deteriorating. If there's too much humidity, some veggies become breeding grounds for molds and bacteria. To control humidity within a space, you need a means of monitoring relative humidity and a way to add or remove moisture from the air.

Hygrometers are to humidity what thermometers are to temperature: They monitor and display the relative humidity of surrounding air. Likewise, humidistats are the humidity analog to thermostats that allow you to control electrical equipment based on relative humidity. It's great to have measurements of relative humidity, but how do you know they're accurate? While InkBird claims that its digital humidity controllers have an accuracy of +/− 3 percent up to 99 percent relative humidity, some people question the accuracy of digital hygrometers in general when relative humidity is above 85 percent.[5]

There are a couple of ways to verify that your hygrometers are reading accurately. First, you can calibrate them, and there are two common and easy methods. One is the damp-towel method, which involves sealing the sensor probe and a saturated cloth or paper towel together in a plastic bag for at least an hour. After an hour, the relative humidity inside the bag should be at (or very near) 100 percent. The second calibration method involves saturating a small amount of table salt (several tablespoons is enough) with water, and sealing that inside a container with the sensor probe. After about an hour, the relative humidity should read exactly 75 percent. Another way to test your hygrometer is to compare it against another measuring tool. A sling psychrometer can produce accurate readings even at very high relative humidities. These are simple to use and consist of two thermometers: one that's dry (the dry bulb) and one that's covered in a wet wicking material (the wet bulb). After noting both temperatures, you can use a psychometric chart or table to determine the relative humidity. Chris Callahan, at the University of Vermont Agricultural Engineering Extension, created a

A commercial-grade centrifugal humidifier connected to a digital humidity controller is an effective way to maintain and control high levels of humidity in a cold-storage room.

Dehumidifiers (bottom left) can reduce the relative humidity for crops that prefer drier conditions in storage. *Photo courtesy of Holly Simpson-Baldwin.*

handy psychometric table for temperatures and humidities relevant to high-humidity cold storage for vegetables (see resources, page 237).

One of the most common problems in cold storage for vegetables is low humidity. Even if you're trying to humidify a space, there are factors fighting against you, namely the dehumidifying effects of evaporators and dry air entering the storage room from outside. To keep relative humidity high, you need to effectively add moisture to the air. Some farmers wet the floor with water, but this practice requires frequent attention, is difficult to control accurately, and is frowned upon by the food-safety community. The most effective way to increase humidity is to use a humidifier, and specifically a commercial-grade centrifugal

humidifier. These can put lots of moisture into the air without adding lots of heat to the room, even when the humidity approaches saturation. They are usually designed to hook into a water supply, or they can be gravity-fed from reservoirs. Some models have their own built-in digital humidistats, while others can plug into humidity controllers for accurate control. You can use the inexpensive wicking-type humidifiers for your warmer and drier storage spaces, but they struggle to increase humidity in temperatures near freezing and at relative humidities above 85 to 90 percent.

There are plenty of situations in which you'll want to reduce the humidity in a space. I dehumidify during curing and the initial storage period for my winter squash and pumpkins. I've found that residential dehumidifiers meet my needs, but depending on your scale of production, you might need to upgrade to larger commercial dehumidifiers if the smaller units aren't cutting it. Two things to note about dehumidifiers: They will produce both water and heat. Obviously, the moisture they pull from the air has to go somewhere, and while most small dehumidifiers have reservoirs, most units also have hose/drain hookups. With a drain hose, you can trust the unit to run continuously rather than stopping when its reservoir is full. Dehumidifiers will also add heat to any space they're in. That's because they work similarly to cooling systems and rely upon cold coils to condense moisture from the air. This presents some possible problems for dehumidifying at low temperatures, since the fins and coils will readily ice up if the air in the room cannot defrost them effectively. Thankfully, many commercial-grade dehumidifiers have automatic defrost settings that allow you to use them in cold spaces.

Another method for removing humidity from a space is to introduce relatively dry air. Anytime you introduce dry air into a space—either from opening the door or using fans—you dilute the humidity inside whether you intend to or not. This is something I need to fight against at Offbeet Farm since I cool my storage room with outside air for most of the winter. The more outside air coming in, the more humidity I need to replace. At Food Farm in Minnesota, they use outside air and a differential thermostat to affect humidity in their potato storage. At the beginning of the storage season, potatoes are usually wet from washing, so they set the differential to 0°F. This means the intake fan runs more often, introducing more dry air to remove excess humidity around the potatoes. After a couple months, they set the differential much higher. This means only relatively cold air enters the storage room in smaller quantities for cooling, which removes less humidity from the room.

Airflow and Fresh Air

Most farmers want their climate-controlled conditions to be consistent throughout their storage rooms,

and good airflow is the key to achieving this. Air movement over, under, and around your storage containers will reduce cold spots in the depths of winter and help to quickly remove field heat during harvest. Creating circular airflow is one effective strategy for ensuring even storage conditions. If, for example, your evaporator fans blow air in one direction at ceiling height, place fans at floor level on the opposite side of the room to blow air back toward the evaporators and create a circulation pattern. The air needs channels in which to flow, so it's good practice to leave space between containers and the ceiling/walls,

Using Uneven Conditions to Your Advantage

Some farms have a hard time maintaining even conditions in their storage rooms despite good airflow. This is especially true with large or narrow rooms. Goranson Farm near Dresden, Maine, uses uneven conditions in their storage room to meet the needs of different crops. Their cold-storage room is long, measuring roughly 30 by 100 feet. The evaporators are on one side of the cooler, along one of the short walls. Despite additional ventilation fans, a temperature gradient naturally forms in the room, with the lowest temps closer to the evaporators. This occurs partly because the thermostats are located behind the evaporators and partly because the room is just so long. The farm uses this temperature gradient to their advantage, storing crops that prefer colder storage conditions—such as carrots and beets—closer to the evaporators where the temperature stays below 34°F. They keep potatoes on the other end of the cooler where temperatures are closer to 36°F.

as well as between the containers themselves. Be sure to orient pallet openings in the direction of airflow as well (see "Airflow, Access, and Future Growth," page 103).

Bringing fresh air into the storage room is also important for maintaining vegetable quality in storage. High carbon dioxide concentrations can harm veggies, and certain crops deteriorate in the presence of ethylene (see crop-specific information in part 3, page 123).[6] Another reason to add fresh air to your storage room(s) is to help reduce pathogen loads in the air. I heard an anecdote about an HVAC installer who regularly worked with farms. They reportedly said that diluting the concentration of pathogens in the air reduces storage rots in industrial potato storage. I haven't found any evidence from research to back this up, but it makes logical sense. Reducing the concentration of mold spores in the air sounds like a reasonable way to reduce the likelihood of some storage rots. As mentioned earlier, cooling with fresh outdoor air is an option for farms in cold climates. Fresh air also naturally enters the storage room as you open and close the doors. If you're not regularly introducing fresh air to your storage rooms, consider adding some periodic ventilation. There are even options for retaining heat and humidity in ventilated storage spaces with ductless HRVs (heat recovery ventilation) or ERVs (energy recovery ventilation).

A last note: Some industrial-scale farms also manipulate the composition of the air in storage, usually adjusting levels of oxygen and carbon dioxide. While some crops can benefit from controlled-atmosphere (CA) storage—Brussels sprouts, cabbages, garlic, onions, sweet potatoes, and winter squash, to name a few[7]—most other storage crops don't. The benefits of CA storage are relatively slight to justify the expense on most small farms, and I've never met a small-scale vegetable farmer using it. That said, it might be something to consider if you specialize in one of the crops that benefits from CA storage.

Part 3
Storage-Crop Compendium

One of my main motivations for writing this book was a longing for the information in this chapter. Successful storage farming requires an additional level of know-how beyond growing the crops, and I so often yearned for not only the "what" but also the "how" and the "why" while learning the ropes. And let's face it: There's a lot of bad information out there. The internet, for better or worse, has democratized the dissemination of information. For subjects as popular and complicated as gardening, there are a lot of people writing a lot of things—some good, some bad, some inspiring, and some dismally misinformed. While researching for this chapter, I waded through books, scientific journals, and the deep marshes of the internet to find what works and why. Problem is, that's not always clear when there's conflicting information. When presented with conflicts, I usually give more weight to scientific trials and studies because they attempt to remove subjectivity and isolate variables. Scientific works also tend to describe the full set of circumstances leading to the specific results. That said, I also trust the experiences of farmers, even when they differ from scientific results. (Farmers are scientists-of-a-sort, after all.) As such, this chapter draws heavily upon my own experiences and those of other farmers in addition to published science. The more I've learned, the more I've realized that the world is complicated and that different methods work in different contexts. When faced with legitimate contradictions, I've tried to present the differing opinions and techniques.

One goal here is to provide the practical information you need to successfully choose, harvest, and store winter vegetables. Each vegetable has information on choosing varieties, harvesting, trimming and processing, curing (if necessary), optimal storage conditions, responses to ethylene (if any), and common problems. It's important to note that fast postharvest cooling will benefit all the crops stored in cold and damp conditions; apart from providing the right storage conditions, this is one of the best things you can do to reduce storage losses. I encourage you to seek out my references for more detailed information. The USDA's *Commercial Storage of Fruits, Vegetables, and Florist and Nursery Stocks* is a particularly excellent resource, but it doesn't always include the full story or practical details relevant to small-scale farmers. I hope this chapter fills those gaps and alleviates anxiety about growing and storing your own vegetables throughout the winter.

Beets

Beets (*Beta vulgaris*) are members of the goosefoot family (Chenopodiaceae), along with spinach, quinoa, and purslane. Originally cultivated by the Romans as a leafy vegetable, this plant gradually morphed into various forms, including Swiss chard, fodder beets, sugar beets, and common table beets.

Beets get their deep reds, pinks, and golds from pigments called betalains, and their earthy (some might say dirt-like) flavor from aromatic geosmins. Many farmers use them as forage crops for animal feed because they're packed with sugars. Those concentrated sugars also inspired the French to develop sugar

Photo courtesy of Phil Knapp.

beets as an alternative to sugarcane while under English blockade during the Napoleonic Wars.[1]

Most customers recognize beets as staple winter storage crops, but there are many people out there who don't like their earthy flavor. Customers regularly receive beets in my CSA, but I try to include some less-intense golden and white varieties. I also find that market customers are drawn to the vibrant and contrasting colors of different beet varieties.

Variety Selection

Some farmers I interviewed didn't worry much about choosing beet varieties for storage, noting that many varieties store well. I spoke to University of Wisconsin professor and beet breeder Irwin Goldman, who noted that some varieties store better than others. He speculated that the thickness of the cuticle (skin) and the size of the crown may affect storability. This might be because varieties with thicker, tougher skins better resist nicks during harvest and processing that potentially become infected with pathogens in storage. On my farm, I've also noticed that varieties with lots of fine roots, such as Chioggia and white varieties, are quicker to sprout in storage. Skin blemishes are also more apparent on lighter varieties than on dark red beets. That said, most beet varieties store up to four months without issue.

Cold Hardiness

Although beets can handle light frosts to their leaves, I protect them from harder frosts and frost on their root surfaces. Although it isn't always immediately apparent, light tissue damage to the outer root surface provides an easy foothold for mold growth. Varieties with thick leaf canopies, such as Boldor, offer good protection to the roots below. If light frost threatens, I often cover varieties with sparser leaf coverage, such as Subeto, to avoid damage on the root crowns and shoulders.

Harvest Indicators

Beetroots are ready to harvest and store at almost any size. I've seen tiny, marble-sized beets store just as

long as their larger counterparts. The question is, what do you and your customers want? Most people prefer beets at least 1.5 inches (4 cm) in diameter, and, of course, larger beets provide more to eat and more to sell.

Trimming and Processing

Some farmers rip beet greens off by hand, but I prefer to cut the tops, if not for storage quality, then for aesthetics. My goal is to remove as much as possible without cutting into the root itself. The outlines of the individual leaf bases should still be visible after cutting. I typically make one horizontal cut, but some farmers make angled cuts following the curvature of the crown to keep the remaining petioles uniform in length.

Beets can be washed prior to storage but should be allowed to dry if you're using enclosed storage containers. Some varieties are easy to bruise or nick during washing, so be careful to avoid damage.

Storage Conditions

Beets do best in cold and damp storage conditions, ideally at 32°F (0°C) and 98 percent relative humidity (RH). At ideal humidity, storage containers can be open to the air. If the storage room air is drier than 98 percent RH, then the containers should be enclosed but allow enough air exchange to prevent condensation. A good storage variety should last between four and six months in the right conditions.

Common Problems

Beets sprout toward the end of their storage lives by growing new leaf shoots and new roots. They may also decay, usually in the form of black spots or softening that often start in the crown and spread inward. This type of decay is easy to spot in light-colored varieties, but you'll need to carefully inspect dark red beets. Gray mold from *Botrytis cinerea* is also common on areas of root damage. Both black rot (*Neocamarosporium betae*) and white mold (*Sclerotinia* spp.) can damage roots if air circulation is poor.[2]

Brussels Sprouts

Brussels sprouts are members of the highly adaptable species *Brassica oleracea*. While they don't have the same fortitude in storage as their close relatives, cabbages, this increasingly popular fall vegetable can last for a month or more if stored properly. One reason for Brussels sprouts' increasing popularity is that scientists recently identified the compounds responsible for their sometimes-bitter flavor, and breeders created new hybrids of older, less-bitter varieties and newer, high-yielding varieties. Chefs caught word and have, in part, driven a Brussels

Photo courtesy of Zoe Fuller.

sprout resurgence with new, trendy dishes such as pan-seared Brussels sprouts and bacon.[3]

Variety Selection

Brussels sprouts have relatively short storage lives regardless of variety, but slow-growing and cold-hardy varieties will allow you to leave the plants in the field longer without damage or over-maturation.

Cold Hardiness

Brussels sprouts are very cold-hardy vegetables. Exposure to light frost improves their flavor as they produce more sugars to prevent freeze damage, but the sprouts will freeze if exposed to prolonged temperatures below 30°F (-1°C).[4] Sprouts can generally handle a few light freezes, but you should harvest or protect them before the onset of temperatures that will turn them into little ice marbles.[5]

Harvest Indicators

The ideal size of Brussels sprouts is more a matter of preference than of storage life. Most consumers expect Brussels sprouts ranging from 1 to 2 inches (2.5 to 5 cm) in diameter. Sprouts mature from the bottom up, but topping plants (removing the terminal bud) when the bottom sprouts measure about 1 inch can help achieve more uniform size across the stalk. Avoid overmature sprouts, which form looser leaves as they enlarge. Typically, the colder temps and shorter days of fall slow growth such that overmature sprouts aren't a problem, but you can cut back leaves and twist the entire plants (to break some, but not all, roots) if you're worried.

Trimming and Processing

Depending on your space limitations and preferences, you can either cut sprouts from the stems at harvest and store the loose sprouts, or you can harvest and store the stems, sprouts and all. Some customers appreciate receiving the whole stems and use the stalks for making soup stock or grilling, while others just want the sprouts.

Storage Conditions

Brussels sprouts store longest when kept near 32°F, and they need 95 percent RH or higher to prevent drying out. Loose sprouts are more prone to drying out but reliably last for three to five weeks. Keeping sprouts on the stalks will buy a few additional weeks in storage. Good ventilation is especially important to avoid condensation and early yellowing of leaves.

Ethylene

Brussels sprouts produce very small amounts of ethylene, but they are sensitive to ethylene exposure, showing damage in the form of yellowing outer leaves that eventually fall off. This is one reason to ensure good ventilation in storage, since the sprouts produce more ethylene gas the longer they're stored.

Common Problems

You'll lose the most salable Brussels sprouts from yellowing leaves. Good ventilation will help the plants continue respiring and prevent the buildup of ethylene and carbon dioxide. The most likely places for decay are the cut stems or the cut ends of the sprouts themselves.

Cabbage

Cabbages are royalty amongst storage vegetables—Brassicaceae is the cabbage family, after all—and they are staple crops on most storage farms. They are relatively easy to grow, and good storage varieties last for remarkably long periods in the right conditions. In fact, cabbage kept much of the population of medieval Europe fed during winter months. During that time, the upper classes considered cabbages too commonplace to grace their tables, and cabbages were thought to impart both melancholy and nightmares upon those who ate them.[6] Nowadays, people think of cabbages as superfoods. A trait common among cultures with high numbers of centenarians is diets rich in cabbage (and other cruciferous vegetables).[7] As a frequent cabbage-eater myself, I gladly accept the risks to my mood and sleep.

Variety Selection

When it comes to storage, not all cabbage varieties were created equal. In my experience, varieties that mature quickly in the field tend to deteriorate more quickly in storage than slower-growing varieties. While green and red storage varieties reliably last all winter, round and pointed varieties meant for early marketing may only last a few months. I've seen certain savoys last four to six months, while the storage lives of napas are considerably shorter. When in doubt, stick to storage varieties, and trial anything else you're interested in growing for storage (see "Variety Trials for Storage," page 10). It may be tempting to grow the big, juicy early varieties, but watching them turn into a slimy mess after a month or two in storage is disappointing. It's also difficult to get the smell of

Photo courtesy of Phil Knapp.

129

disappointment out from under your fingernails after a couple hours spent peeling rotting cabbages.

Cold Hardiness

Cabbages can handle some light frosts, but their outer layers readily freeze if exposed to below-freezing temps for longer periods. This doesn't always spell disaster, since partially frozen cabbage leaves sometimes bounce back after a single freeze, but you should avoid exposing them to multiple freeze-thaw cycles. At worst, you can peel away the damaged leaves before storage. (Previously frozen leaves usually turn into goo at some point during winter storage.) I've observed that tender, inner cabbage leaves readily freeze while transporting cabbages to markets during cold winter conditions. The freeze damage appears as spotting and eventually widespread water-soaked tissue. The longevity of the damage depends on how extensively the tissue was frozen; sometimes the damage goes away, while more often, the damaged layers soften, and I need to peel them.

Harvest Indicators

From the perspective of storage life, harvesting immature cabbage isn't a problem, whereas harvesting overmature cabbage is. Small, loose, unformed heads will last nearly as long in storage as larger, tight heads of closely packed leaves. Cabbages that have overmatured and split, picked up fungal or bacterial problems, or (god forbid) bolted will not last long in storage. Unless they're making sauerkraut, most customers prefer tight small-to-medium heads.

Trimming and Processing

Wrapper leaves—the thick, darker pigmented leaves that curl away from cabbage heads at their tips—protect cabbages from drying out in storage. Once cut from the main stalk, you can trim cabbage butts to leave the final set of desired wrapper leaves in place.

Farmers differ on how many wrapper leaves to keep, but all the farmers I interviewed leave at least one set. The drier your storage space, the more wrapper leaves you should leave on, understanding that these extra leaves create more bulk and weight to haul around. When you take the cabbage out of storage, trim the butt to remove all wilted leaf layers and expose the pristine head beneath.

Storage Conditions

Cabbages keep best at 32°F and 98 percent RH. Any colder and the heads are liable to freeze, whereas warmer temps will encourage early sprouting and accelerate problems with decay. The outer layers tend to get slimy in storage if kept in enclosed containers (even with perforations), so many farmers choose to keep cabbages in open bins or piles, even if humidity is lower than ideal. (Below the ideal humidity, the outer layers of cabbages will wilt over time and should be peeled away while processing for sales.)

Ethylene

Cabbages produce very little ethylene but tend to yellow in the outer leaf layers when exposed to the gas.

Common Problems

Over time, the cut stems turn dark and might show signs of sprouting; retrimming prior to sale solves this and makes cabbages look freshly harvested. As cabbages age, their outermost leaves turn yellow. Occasionally, you'll see grey molds (*Botrytis* fungi) form on the outer layers as they deteriorate or on wounds.[8] If there's not enough airflow, these issues usually turn slimy but can be removed by peeling back several layers, which usually prevents rot from spreading further inward. It's common for the leaf tips to senesce together toward the end of a cabbage's storage life. If this happens, there will be off-color leaves throughout the heads, usually making them unsalable.

Carrots

Although wild carrots (*Daucus carota*), commonly called Queen Anne's lace, can be found throughout the world, domesticated carrots (also *Daucus carota*) likely arose in two locations: the Iranian plateau and modern-day Turkey. Roots of the more-eastern variant tended to be purple, red, and white, whereas the more-western variant tended toward oranges and yellows.[9] It's difficult to track how these early cultivars combined to create modern-day carrots because, at the time, the people who wrote things down readily confused carrots with parsnips.

Carrots are shining stars on most storage farms. As Steve Pincus from Tipi Produce put it, "You can sell carrots almost anywhere in the world in winter because people *want* them." I wholeheartedly agree with his sentiment. Locally-grown carrots taste markedly better than store-bought carrots from far away, and this difference is one way to get people excited about local storage crops in general.

Variety Selection

Without exception, every storage farmer I've spoken to grows Bolero as their main storage carrot. Bolero is an orange hybrid, Nantes-type—named for the city of Nantes, France, where botanist Henri Vilmorin developed them in the 1850s—bred by the French company Vilmorin. Unfortunately for certified-organic growers, Vilmorin does not offer an organic option, and you'll be hard-pressed to find truly organic Bolero seeds.

For organic growers and those looking to branch out beyond Bolero, most carrot varieties will store for at least three months without issue. As with cabbages, I find that fast-maturing carrots tend to perform poorly in storage compared to varieties with more days-to-maturity. Look for varieties that grow vigorously in your local climate, cope well with local disease pressures, and tend not to split or break.

Cold Hardiness

Carrots can be exceptionally cold hardy if the conditions are right. Issues arise when there are prolonged below-freezing temperatures. First, the carrot crowns poking out of the ground are liable to freeze. Once frozen and thawed, carrot tissue often shows cracking

Photo courtesy of Phil Knapp.

and takes on a spongy texture that molds quickly in storage. Second, and more annoyingly, the soil surface freezes, creating a crust that's hard to get through and sticks to the carrots. You can mitigate these problems by mulching your carrots—straw, tarps, and blankets all work. The belowground portions of carrot roots are remarkably freeze-resistant. I've found carrots in the spring that made it through an Alaskan winter relatively unscathed. Out of the ground, carrots are very susceptible to freezing damage. If hard frosts are coming, I highly recommend either harvesting or covering your carrots.

Harvest Indicators

Immature, pencil-thin carrots tend to dry out or deteriorate quickly in storage because their periderms—the outermost root surfaces containing protective waxes—haven't fully developed. Maturity in carrots seems to coincide with root tips filling out. This isn't always obvious, especially for inherently pointy carrots such as Imperator-types. Even small carrots can be mature enough for storage. Just make sure they're approaching the right shape for their type, be that conical or cylindrical.

Trimming and Processing

Carrot tops should be trimmed down to, at most, half an inch (1.3 cm), and preferably right down to the crown. Some farmers prefer to rip off tops while others prefer to cut tops with a knife; it comes down to preference, aesthetics, and speed. (Some varieties with beefy tops, such as Yellowstone, are very difficult to rip off cleanly.) Excess or jaggedly cut tops sometimes mold in storage, but small fragments of greens usually don't cause problems.

You can store carrots unwashed or washed. Unwashed carrots might have soil stains if kept more than a couple months, especially with clay soils. Pressure washers and brush washers can sometimes remove stains. Washed carrots should be allowed to air dry before or during storage to avoid mold growth.

Storage Conditions

Carrots store best at 32°F and between 98 and 100 percent RH. With adequate humidity, carrots can be stored in fully open containers (or in piles). However, if the storage room is too dry, partially closed containers with enough ventilation to avoid condensation can keep carrots turgid. Many farmers place polyethylene bin liners upside down over the tops of bulk containers, leaving the bottoms open for ventilation. Whatever you do, don't store carrots in fully sealed containers that don't allow air exchange; the carrots will turn to goo at some point.

Ethylene

Carrots don't produce much ethylene in storage, but exposure to ethylene gas produces bitter-tasting compounds in their skins.

Common Problems

One of the most common storage problems with carrots is drying out. (This said, floppy, desiccated carrots are some of the best-tasting carrots you'll ever eat. Maybe someday I'll convince the world of this.) If your carrots are floppy, the humidity is too low. If you're using partially enclosed containers, it's common to have a few carrots dry out near the ventilation openings. On the flip side, too much moisture and condensation often lead to white mold growth. Inadequate airflow encourages both condensation and growth of fungal pathogens. Some of the most insidious are in the genus *Sclerotinia*, which cause damaging white molds in storage if populations are allowed to build up in the soil.

Toward the end of their storage lives, carrots begin growing root hairs and sprouting new leaves. I've found it's possible to hamper root growth by lowering the humidity to dry out the new growth (not too low as to dry out the carrots themselves). To a point, you can scrub off root hairs and cut shoot growth, but after a while it becomes a losing battle. The carrots' flavor also suffers as sprouting advances.

Celeriac

Celeriac (*Apium graveolens* var. *rapaceum*) is counted in the same species as celery, whose greenish-white stalks adorn our vegetable trays and flavor our soups. Also known as celery root or turnip-rooted celery, celeriac brings the familiar flavors of celery in an enlarged, fleshy root (actually, hypocotyl) that can store for months in the right conditions. Celeriac was developed from wild celery (same species), which grows naturally throughout the Mediterranean Basin but wasn't developed or bred widely until the nineteenth century.[10] Because it is an uncommon and weird-looking storage crop, you might need to provide some education to customers. Celeriac provides that classic celery flavor to soups and stews, but it's also tasty raw grated into salads.

Variety Selection

One trait to pay attention to when selecting varieties is uprightness, or how much the round root/hypocotyl sticks up above the soil. The further the celeriac sits

Photo courtesy of Amy Frye.

underground, the more small and fine roots you'll have to trim off, since those roots trap lots of soils and aren't usually eaten. But the further the hypocotyl sits underground, the more it's protected from cold temperatures. By harvesting earlier and avoiding cold weather, you can choose more upright varieties with fewer roots and cleaning.

Cold Hardiness

Light frosts will improve celeriac's flavor, but prolonged temps below 30°F (−1°C) or more than a few minutes below 20°F (−7°C) will damage the aboveground portions, especially the celery-like stalks. Such damage can propagate decay into the root tissue if frozen and brought into storage.[11]

Harvest Indicators

Size is the main indicator of maturity. Preferably, celeriac are as large as possible without bolting or splitting, but small celeriac will store just as well. Customers usually want celeriac baseball-sized or larger.

Trimming and Processing

Farmers usually trim both the leaves and the small roots off celeriac. Because the bottom root mass is so thick—people often compare it to Medusa's twisted head of snake-hair—it traps a lot of soil and needs to be removed lest one get dirt in their dinner. Trim the leaf stalks to between a half and full inch (1.3 to 2.5 cm) before storage, and trim the roots to expose all the nooks and crannies before washing. You might want to leave room for further trimming to make the celeriac more attractive before selling.

Storage Conditions

Celeriac like cold and damp storage conditions, at 32°F and at or above 98 percent RH. In my experience, the more airflow the better. It's easy for moisture to get trapped within the twisted root mass, even if it's trimmed. If you've been aggressive in your trimming, err on the humid side to limit moisture loss, whereas drier conditions will help remove excess moisture from more-intact leaf stalks and roots to limit decay.

Ethylene

Celeriac don't produce much ethylene in storage. If exposed to ethylene, especially at warmer temps, celeriac tends to lose green coloring on its root shoulders and leaves.[12]

Common Problems

As members of the parsley family (Apiaceae), celeriac are susceptible to the same storage diseases that commonly affect carrots and parsnips. Airflow is extremely important; containers that are too tight will quickly lead to molds and rots.

Daikon and Korean Radishes

In 2005, residents of the Japanese city of Aioi found a daikon radish that somehow germinated in a crack in the asphalt, eventually breaking through and growing into a vigorous plant. People noticed the unlikely vegetable and, inspired by its determination and perseverance, dubbed it *dokonjo daikon*, loosely meaning "the radish with the fighting spirit." The dokonjo daikon became something of a local celebrity, and city officials erected a sign to protect the radish and encourage people to treat it with respect. At some point, local hooligans decapitated the famed radish but were so admonished by the backlash that they returned its cut remains to city hall. City officials rushed the radish to a local agricultural research station, where staff managed to propagate seeds from the cut pieces. Now others can grow and admire their own piece of cruciferous resilience.[13]

Daikons and the other winter radishes are indeed resilient. The information in this section refers to all winter radishes—including daikon, Korean, and black radishes—which are highly similar when it comes to cultivation and storage. These varieties are

Photo courtesy of Amy Frye.

all members of the same species, *Raphanus sativus*, as the European radishes that fill spring markets and tables. Winter radishes are uniformly larger and slower growing than their spring siblings; most take twice as long or more to mature. This slower maturation time is one reason why winter radishes store so well. They are particularly popular among customers who dabble in fermentation, because they are essential ingredients in kimchi.

Variety Selection

In my opinion, radish variety selection is more about their end use than their growth or storage traits. Pay attention to how much they stick up out of the soil and to the fullness of their foliage. These will be factors that affect how much cold the radishes can tolerate before harvest: The more foliage and the less root shoulders poke up above the soil, the hardier plants are against cold temperatures.

Cold Hardiness

Winter radishes are hardy to light frosts. Because so much of their root is above the soil, they cannot tolerate prolonged temps below 28°F (−2°C) or anything below 20°F (−7°C) without sustaining damage. They may bounce back from one light freeze, but multiple events will damage them. Freeze damage usually manifests as water-soaked tissue that quickly shrivels and deteriorates in storage.

Harvest Indicators

Winter radishes are another vegetable that can be harvested at any size for storage. Larger roots are less likely to dry out, but you can harvest anything that's reached marketable size. Most customers appreciate round winter radishes larger than gumballs and cylindrical radishes at least 1 inch (2.5 cm) wide at the crown. Radishes are prone to bolting, so be sure to time plantings and harvest to avoid overmaturation.

Trimming and Processing

Trim tops down to a quarter-inch (0.6 cm) but without cutting into the root flesh.

Storage Conditions

Radishes do best at 32°F and above 95 percent RH. They are not as susceptible to drying as carrots and parsnips but will become soft if the humidity is too low. Containers that allow small amounts of airflow are good choices. Winter radishes usually last between two and four months in good storage conditions, though some varieties may last even longer.

Ethylene

Radishes produce small amounts of ethylene but are not susceptible to exposure unless the tops are present. Radish tops exposed to ethylene will quickly yellow.[14]

Common Problems

Radishes are favorites of cabbage root maggots (*Delia radicum*), and maggot-damaged root tissue, apart from being gross, deteriorates quickly in storage. If trimmed, the cut surfaces tend to dry out and will shorten the storage lives of the affected radishes.

Garlic

Garlic is hugely popular among my customers, and it's likely the local market is hungry for it no matter where you live. While researching, I came across an interesting and surprising fact: Until recently, all cultivated garlics (*Allium sativum*) were sterile. Thus, the only means of growing more garlic was vegetative, using cloves (or bulbils from hardneck garlic) to grow the next crop. Humans have been breeding garlic for more than 10,000 years. During that time, we encouraged garlic to put so much energy toward bulb growth that it eventually lost the ability to grow viable seeds.[15] This presents some problems for adapting to new conditions and eliminating viruses that build up in a seed stock. However, in 1983, a Japanese researcher reported success growing fertile garlic seeds produced from self-pollinated plants found in Soviet Russia.[16] This discovery opened the door for the possibility of crossbreeding, but we've yet to see commercially available garlic hybrids.[17]

Variety Selection

There are oodles of garlic varieties to choose from, but they are usually broken into two general groups: hardneck and softneck. You'll sometimes hear these referred to as bolting and nonbolting. These secondary names are worth knowing, because they describe the reason for the more common names: Hardneck garlics bolt, sending forth the hard flowering stalks known as scapes, whereas softneck garlics don't bolt and retain soft necks like onions. But hardnecks won't bolt naturally in all climates. Hardneck garlics need exposure to a month or more of temperatures below 40°F (4°C) to send forth their scapes. As a sidenote, I once thought the term *scapes* alluded to the flowering stalks e*scap*ing from the plant in some manner. It's actually a botanical term for a long *peduncle*—or flowering stalk—arising from an abbreviated stem.

For many farmers, geography and climate limit their variety options. Softneck garlics grow best in warmer regions, producing larger bulbs with more cloves. They store for a long time, up to twelve months in some cases, but they have difficulty overwintering in colder climates. That said, some softneck varieties, such as Inchelium Red, are adapted down to USDA hardiness zone 3. Hardneck garlics produce smaller bulbs but fewer and larger cloves compared to softnecks. They are better adapted to growing in cold climates and are the primary type grown in the northern United States and Canada. Hardnecks don't store as long—usually up to six months—but provide the added benefit of scapes, the delicious curlicue

Photo courtesy of Phil Knapp.

Growing Garlic in a Temperate Rainforest

Farragut Farm is located in a coastal temperate rainforest about 30 miles (48 km) by boat from Petersburg, Alaska. A few years back, I had the

opportunity to work several weeks on Farragut Farm after becoming friends with co-owners Bo Varsano and Marja Smets. While I was there, they

Garlic drying on racks in a Farragut Farm greenhouse. *Photo courtesy of Marja Smets.*

flowering stalks that emerge in midsummer and must be trimmed to redirect growth to the bulbs.

Cold Hardiness

Cold-hardiness concerns are different for garlic than for other storage crops. Most growers in cold climates harvest garlic during the summer and plant in the fall for overwintering, so the main concern is whether it can survive in the ground during the winter months. When purchasing seed garlic, pay attention to each variety's hardiness rating (the lowest USDA hardiness zone in which it can survive reliably). Know, however, that overwinter survival is complicated. Low temperatures, planting depth, snow cover, soil moisture,

and mulch can all influence whether or not those garlic cloves survive. Interestingly, garlic cloves can supercool well below their freezing point without the formation of damaging ice crystals—as low as 6°F (−14°C)![18] The soil around garlic cloves can thoroughly freeze without killing them.[19] Heavy mulching can help alleviate freeze-thaw cycles that might damage the garlic and also provide insulation for varieties that might not otherwise survive in your climate.

Harvest Indicators

Try to harvest the largest bulbs possible without them splitting or decaying. One good measure of maturity is leaf dieback. When half the leaves, starting from the grew garlic for market alongside a diverse offering of fresh veggies, but they've since shifted to focus on a few crops that do particularly well, including garlic. I have a hard time thinking of a more difficult climate for curing garlic, but they've made it work. If they can cure garlic successfully in the cool, damp conditions at Farragut Farm—which is fully off-grid—then you can too!

When they doubled down on garlic, Marja and Bo also reevaluated their system for harvesting and curing. Before, they hung garlic in bunches from the rafters of their shop and blasted them with air using box fans. This process was time-consuming, and sometimes the garlic struggled to dry and showed mildew spotting on the outermost layers. Now they do a lot of trimming and cleanup in the field during harvest. They remove most of the stem and foliage to leave behind a 6-inch stub, clip the roots nearly to the basal plate, and peel off the outermost dirty skin layer. Then, they put the garlic stems upside down through 1-inch hardware cloth stretched over wooden racks stacked in their seed-starting greenhouse (which is outfitted with row cover to block direct sunlight). Removing the excess foliage helps them eliminate a lot of excess moisture and biomass that previously hampered air movement and drying. Peeling off the dirty skin layer eliminates any need to clean the bulbs later, so they avoid a washing step, the prospect of which made them cringe. Any water that's managed to get down the cut stem drains out once the garlics are upside down in the racks. Using that particular greenhouse allows them to cure garlic in a way that makes sense for their climate and infrastructure. On the rare sunny days, there's enough excess power from the solar panels to run fans on the garlic racks, whereas they're able to light the greenhouse's woodstove on cloudy and damp days to raise the temperature and promote drying. After about a month, the garlic bulbs are fully dry. Bo and Marja don't sell much garlic from storage, saying they market most of their crop right away, but they and their customers report great results storing the garlic at home.

bottom, have turned yellow, the garlic is likely ready. Dig up a couple sacrificial test plants and check the bulb shoulders, which should be definite and pronounced. If the bulbs look like torpedoes, delicately and evenly tapering to the stems, they're not ready. If you cut the bulbs in half, you should see cloves fully filling their wrappers and, with hardneck garlic, pulling away from the flowering stalk. Most experienced growers recommend cutting off irrigation for one to two weeks before harvest to encourage dry-down and avoid decay.

Trimming and Processing

There's some debate about whether trimming garlic before or after curing is best. Some growers claim that the full leaves are necessary for quickly pulling moisture from the bulbs' outer scales. Others assert that the extra moisture and humidity from the tops either slows down the drying process or has no effect. When I started growing garlic, I hung and dried whole plants with good results. But later, I attended a presentation by Cindy Hollenbeck of Keene Organic Garlic near Madison, Wisconsin, where she described their method of mowing garlic stems down to 6 inches (15.2 cm) before harvesting. I was surprised but intrigued, and I tried it the next season with great results. Now I prefer the trim-first method because it saves both time and space. Without the tops, I can cure my garlic in crates rather than hanging them, and most of the trimming is already done. If you're feeling incredulous, read Growing Garlic in a Temperate Rainforest on page 138.

Curing

With garlic, curing dries the outermost scales and the necks to seal in the cloves, protecting them against both moisture loss and pathogens. Even in relatively dry climates, a curing period tends to reduce losses in storage.[20] The process of curing garlic is very similar to that for onions and can happen in a variety of warm and dry conditions over one to several weeks. Temps between 80 and 90°F (27 and 32°C) and about 65 percent RH are ideal. Whatever the conditions, you also need excellent airflow, and most farmers set up fans to force air over the drying crop. It's common to hang garlic bunches to aid drying. At my first farm in Michigan, I used an old swing-set frame in a garage to hang and cure lots of garlic in a small space. You can also cure garlic on racks and in well-vented bins, and many farmers cure their garlic in warm sheds or greenhouses. If curing in the field or in a greenhouse, you'll need to protect garlic from the sun since it's susceptible to sunburn or sun scalding.[21] You can trim garlic necks before curing, but be sure to leave at least 6 inches to prevent pathogens from moving downward through the neck.[22]

Storage Conditions

Garlic stores best in cold and relatively dry conditions: 32°F and 60 to 70 percent relative humidity.

Garlics—like their fellow members of the lily family, onions—go dormant after curing, and this dormancy is controlled via plant-growth hormones that respond to environmental conditions. When triggered, these hormones encourage garlic shoots to grow, at first within individual cloves and eventually bursting forth from their tips. When this happens, the growing shoots start demanding a lot of moisture, and the cloves become soft and shriveled (which is bad for sales). There's a lot of research and experience demonstrating that garlic sprouts quickly between about 40 and 50° F (4 and 10°C).[23] I've read accounts from multiple experienced garlic growers that higher temperatures and lower humidity—around 55°F (13°C) and 50 percent humidity—work well for storage, and I've heard similar things from garlic growers in Michigan.[24] There's also research showing that garlic sprouting is suppressed above 86°F (30°C).[25] My opinion is that very high temperature storage is impractical for most growers. The lowest temperature option is best, but the mid-range option is intriguing because it nearly overlaps with ideal conditions for winter squash and pumpkins.

Common Problems

Getting the humidity wrong in storage can cause problems. If it's too humid, garlic often develop mildew spots or mold growth. High humidity also induces early root sprouting. If the humidity is too low, garlic tends to lose too much moisture, shriveling in the wrappers.

Kale

Similar to the dokonjo daikon (see "Daikon and Korean Radishes," page 135), I once found a kale plant—Russian Red if my memory serves—growing in a heap of sand, rocks, and concrete refuse following a destructive flash-flooding event in Hancock, Michigan, in the Upper Peninsula. I assume a wayward seed got swept along in the deluge and deposited in that pile. Unlike the famous daikon, I don't think this resilient plant received much admiration beyond the few moments I stood and observed, "Huh, look at that." It goes to show, however, that kale is one hardy plant. For a leafy green, kale can withstand shockingly low temperatures and can hold up in storage long enough to supply folks with fresh greens well after winter takes hold. I use kale early in the storage season to give my customers local greens an extra month or two after most other sources have disappeared.

The plants we commonly call kale actually consist of two different species. *Brassica oleracea* var. *acephala* is probably what you picture. These kales are closely related to cabbages. They grow from a thick, central stem and include varieties such as curly kale, Lacinato kale, Scotch kale, and collards. *Brassica napus* var. *pabularia* are more closely related to canola and have shorter, less robust stems than their cousins. Common varieties include Siberian, Russian, Ragged Jack, and asparagus kale.

Variety Selection

When choosing kale varieties, you need to think about your goals. Do you want the kale to maintain excellent quality in the field as long as possible, or do you want kale to last as long as possible in storage? If your climate allows, kale leaves are better off staying in the field attached to the plant. In this case, you want kale varieties that remain beautiful in your climate and put up with your region's low-temperature extremes. The *napus* kales like Siberian and Russian are reported to survive harsher winter conditions than the *oleracea* kales. That said, survival and quality are two different things. In my opinion, kale leaves that have been frozen solid are perfectly edible, but I don't necessarily consider them salable. Ice crystals do funny things to the fleshy parts of leaves, especially the petioles and midribs, making their texture soft and spongy. If you plan to sell previously frozen kale, I'd make sure your customers know that upfront.

If you want kale to last in storage as long as possible, then you want varieties that delay leaf senescence. A recent variety trial from Clemson University reported that Darkibor, Vates, and Curly Roja maintain high quality in storage.[26] In my experience, slight yellowing is harder to see in red varieties, and curly *oleracea* kales and their fluffy leaves facilitate good air flow, but I think the *napus* kales delay senescence the longest in storage. Regardless of variety, the best thing we can do as farmers is to keep the leaves cold. This slows down metabolism and consequently the production of ethylene, the production of ammonia, and the breakdown of chlorophyll. (Ethylene gas triggers chlorophyll breakdown—which leads to yellowing—and ammonia gives kale a bitter flavor.)

Cold Hardiness

Kale is a remarkably cold-hardy vegetable, but that fact should not make you complacent. Despite what people may say, fully grown kale is damaged by more than a few light freezes. This is often fine for the home gardener, and it's a wonderful treat to wade out through the snow to snap off a few frosty kale leaves for dinner. But for the commercial grower, freeze-damaged kale isn't what most customers expect. Kale that's been frozen does not last as long in storage or in the refrigerator, and the texture of the petioles and midribs especially becomes slightly spongy. I wouldn't let it lightly freeze more than a few times. Kale's

Photo courtesy of Phil Knapp.

freezing point is 31°F (−0.6°C), and its leaves are very exposed.[27] If prolonged freezing temperatures are in the forecast—in other words, below 31°F for more than three or four hours—it's time to bring the kale inside unless you're able to protect it.

Harvest Indicators

Although any marketable kale leaves can come into storage, kale leaves react differently to storage based on their maturity. Overmature leaves—those toward the bottom of the plant that have fully expanded and might show slight discoloration—are the quickest to deteriorate in storage, whereas immature leaves toward the top of the plants maintain their color and taste the longest.[28] Since kale leaves release small amounts of ethylene as they deteriorate, you may want to avoid storing overmature leaves alongside the younger kale to avoid early deterioration of the whole lot.

Because kale leaves have relatively short storage lives, it behooves you to leave them in the field as long as possible. I'm thinking about trying a staggered harvest in future years so that only young kale leaves enter storage. When I harvest the entire plant at once, many of the mature and overmature leaves go to waste, when I could have sold them fresh instead.

Trimming and Processing

Most farmers I know pull rather than cut kale leaves off the plants. It's faster and you're less likely to damage other leaves with an errant knife tip flying around. For me, the knife work comes when making bunches for sale. I was taught to align the leaf tips and cut the petioles to an even length. This technique creates aesthetically pleasing bunches and accommodates leaves of varying sizes.

Some farmers wash their kale prior to sale, while others don't. It's up to you and your comfort/obligations regarding food safety. If you choose to wash stored kale, I suggest washing after storage to introduce as little free moisture as possible.

Storage Conditions

Kale does best when kept as cold as possible with very high humidity, preferably 32°F with humidity higher than 98 percent. Kale leaves are very susceptible to drying out, so I don't recommend storage containers with lots of airflow unless you're able to maintain the proper humidity. Many farmers use bags that allow small amounts of air exchange to prevent ethylene and carbon dioxide from accumulating. Though kale's time in storage is relatively short, excess moisture in the form of water droplets can still cause problems. I've seen mold develop around water droplets in a matter of weeks. If healthy kale leaves enter storage in good condition, expect losses to yellowing under 25 percent after four weeks. After eight weeks, 25 to 50 percent of the original crop has usually yellowed. I've seen the youngest leaves last upward of four months without yellowing.

I've tried storing kale a couple of ways. I've ripped leaves off the plants and stored those inside woven poly super sacks. I've also harvested whole plants, roots and all, and stored those in plastic totes with their roots bathed in damp soil. The rates of yellowing were about the same, and I suspect this is partially because the storage room is dark. Even at 95 percent RH, many of the kale leaves in the totes lost too much moisture, despite still being attached to the plants. The kale leaves inside super sacks retain moisture but tend to have more problems with decay and molds, especially when they're tightly packed. I suspect that gases like carbon dioxide and ethylene build up in the tightly packed leaves and hasten their degradation. My best results have been when using super sacks but piling the leaves more loosely—think of a tossed salad. With room to breathe, the leaves had fewer issues with decay and retained their moisture.

Ethylene

As with all leafy greens, kale responds to ethylene exposure with yellowing and further senescence, and the more leaves deteriorate, the more ethylene they produce.[29]

Common Problems

The two most common problems you'll see with stored kales are yellowing and desiccation. Unless you're keeping kale under lights, I don't see a way around the problem of yellowing, although you can mitigate the buildup of ethylene by allowing adequate airflow. The only way to stop kale from drying is to maintain proper humidity, whether that's in the whole room or via a semi-enclosed container. The longer kale stays in storage, the more problems you'll see with decay. Watch carefully for small mold spots that easily hide in the folds of curly kales. Tips of petioles may decay or mold, and midribs eventually yellow. At some point, too much ammonia builds up in the leaves and makes them unpalatable. If in doubt, tear off a piece of leaf and try it yourself—your taste buds won't lie!

Kohlrabi

As a farmer and an eater, I love kohlrabi. From the farming perspective, kohlrabi is a high-yielding crop that's easy to transplant and escapes many of the insect pests that plague the brassica root veggies like radishes and turnips. As an eater, I'm intrigued by their unique appearance and love their crisp sweetness when raw and their buttery smoothness when cooked. However, as my wife constantly reminds me, kohlrabis freak a lot of people out. If there were a frequently asked questions list for winter veggies, "What do I do with kohlrabi?" would be near the top. To start, I usually tell people to think of kohlrabis as gigantic broccoli stems and to substitute them for potatoes. It's the same species as cabbage and broccoli, but in a group all its own, *Brassica oleracea* var. *gongylodes*.

Variety Selection

There are many varieties of kohlrabi, and they don't perform equally in storage. The most well-known storage variety is Kossak. I've had good experiences with these large green kohlrabis, but I've observed issues resulting from their relatively large, fleshy leaf bases. They eventually fall off in storage but can spread rot when deteriorating. I use Kolibri, a purple kohlrabi, as my long-storage variety because, in addition to holding well, its leaf bases are relatively small and more likely to

Photo courtesy of Phil Knapp.

144

dry rather than rot. To date, all the white kohlrabi varieties I've tried have deteriorated quickly in storage.

Cold Hardiness

Kohlrabis are cold-hardy brassicas, but they're also relatively exposed, with their swollen stems sitting entirely aboveground. Unlike cabbages, there's nothing to peel off when they're damaged, so there's less room for error. While they can handle multiple light frosts, harder frosts that threaten to freeze their skin and flesh should be avoided. I recommend harvesting them before they're exposed to prolonged temperatures below their freezing point of roughly 30°F (−1°C).[30] Covering them with row cover—or, better yet, tarps—will buy you some extra time if needed.

Harvest Indicators

Size is the main indicator of maturity, though small kohlrabis store just as well as large ones. If cold weather forces me to harvest, I tend to leave small kohlrabis—anything smaller than a tennis ball—behind. Most people peel off the thick and often stringy skin, so I feel that small kohlrabis are more effort than they're worth for the farmer and the customer. You want to harvest large kohlrabis before they split, but be aware that some varieties tend to crack when they're harvested. Withholding water for a few days before harvest can help alleviate this problem.

Trimming and Processing

At harvest, I trim the small stem on the underside of the kohlrabi to half an inch (1.3 cm) or less. Trim leaves as close to the rounded stem as possible but not so close that you skin the kohlrabi. Sharp cutting shears are nice tools for the job because they're useful for cutting both the woody stems and the delicate leaves. Kohlrabis sometimes have a ground-spot of sorts. Wherever the stem touches soil, adventitious roots form. I usually cut these parts off by skinning that portion of the vegetable to avoid bringing soil-borne pathogens into storage.

In storage, the remaining leaf bases eventually senesce, dry, and fall off after a month or two. If conditions are too moist and stagnant, the bases might turn brown and slimy, and this rot often penetrates and ruins the rest of the kohlrabi. Before sale, I remove any remaining loose leaf bases and wipe down the kohlrabi with a clean damp cloth.

Storage Conditions

Kohlrabis, like the other brassicas, do best when stored in cold and humid conditions—32°F and RH above 95 percent. Kohlrabis are relatively resistant to drying out in storage, and I use my most-ventilated containers to avoid problems with decaying leaf bases. Good storage varieties harvested in healthy condition should last four to five months.

Common Problems

In addition to monitoring the leaf bases as they senesce, you'll want to watch for common storage diseases, including bacterial soft rots caused by bacteria in the genera *Erwinia* and *Pseudomonas*.[31]

Leeks

Leeks (*Allium porrum*), like the other *Alliums*, are cousins to the wild plants that share their common name: *Allium tricoccum* in eastern North America and *Allium ursinum* in Europe and Asia. With all these plants, we're after their succulent leaves, and the underground, blanched leaf bases are their most-prized portions. I fondly recall spring afternoons during my childhood spent digging wild leeks and preparing them using a leek-specific cookbook cleverly titled *First You Take a Leek*.[32] While cultivated leeks taste less garlicky than the wild ones, their delicate flavor, ease of cutting, and surprising hardiness make them popular winter veggies. Perhaps with the exception of some savoy cabbages, leeks are the last dark green leaves many farmers can offer their customers after winter forces everything into storage.

Variety Selection

The best varieties for storage are the "late-season" or "winter" leeks. These are slow-growing, and they often have a blueish tint to their leaves. The earlier varieties aren't as tolerant of cold and won't last as long in storage.

Cold Hardiness

Leeks benefit from our preferences when it comes to cold hardiness. Many people prefer leeks with stalks blanched up to where their leaves fan outward, and hilling is a common technique for achieving this. As such, the portion we're most interested in eating is largely protected from cold temps. The exposed upper leaves are another story, as they can freeze at about 30°F (−1°C).[33] Similar to kale, leeks are not as resistant to cold temperatures as you might think, especially if you're selling them. Although they can tolerate light frosts, leek leaves are damaged when frozen solid, leading to discoloration, early yellowing, and off textures.[34] It's possible to mulch them to prevent freezing damage

to the aboveground leaves, but I wouldn't recommend tarping unless you provide supports to hold tarps off the leeks. Once, I placed a tarp directly over my leeks before a snowfall, and the weight bent them at funny angles. Straw could be an option if you have easy access to it. Otherwise, I recommend harvesting leeks before there are prolonged temperatures below 30°F if you plan to utilize the aboveground leaves.

Harvest Indicators

Unlike the other cultivated *Alliums*, leeks do not enter a state of hormonal dormancy, so size is the only indicator of maturity at harvest time. Most customers prefer leeks at least 1 inch (2.5 cm) in diameter near the base, and I wouldn't market anything smaller than half an inch (1.3 cm). After a time in storage, the necessary peeling and trimming might leave very little to sell and eat.

Trimming and Processing

There's some debate over whether to trim leeks before or after storage. In trials conducted by the USDA, trimmed leeks lasted about a week longer in storage than untrimmed leeks when both were washed upfront.[35] My sense is that this pattern would extend to unwashed leeks, since decay often starts at the leaf tips and subsequently spreads elsewhere.

Trim roots to less than half an inch (1.3 cm). The style and length of leaf trimming are matters of preference; trimming leeks is a rare time for artistic expression! Some farmers trim leek leaves straight across, while others prefer a triangular, fan-like pattern. Some leave long tops, while others trim off most of the greens, leaving only the blanched portion. It's entirely up to you. If you choose to trim before storage, it's a good idea to leave a little extra in case you need to do further aesthetic trimming and peeling after storage.

Photo courtesy of Danielle Knapp.

Note that if you want leeks to have their characteristically blanched stalks—the botanical term for the bundled shaft of leaves is *pseudostem*—you need to hill or collar them early in the season.

Storage Conditions

Regardless of how you store leeks, they like cold and damp conditions—32°F and RH above 95 percent. Based on a set of trials, the USDA recommends storing leeks inside polyethylene bags or liners, either perforated or not.[36] I've also tried storing leeks packed upright into totes (untrimmed) with their roots touching damp soil. Both methods have their merits. In my experience, leeks last longer with their roots in soil. However, this method requires a lengthy cleaning process—made lengthier without easy access to water—and takes extra effort and space to set up containers with soil. If you're using semi-enclosed containers, make sure there's some air exchange to release excess moisture. Depending on how you store them, leeks should last between six and ten weeks before deteriorating beyond what's salable.

Ethylene

Leeks don't produce much ethylene, but they're moderately sensitive to it, showing yellowing leaves and softening with exposure.

Common Problems

Since leeks aren't entirely dormant in storage, they continue to grow. This is especially noticeable if you trim them upfront. Elongation is minimal if leeks are kept at 32°F, but the warmer the temps and the longer they're stored, the more stem growth you'll see.

Onions

Onions have been cultivated for food and medicine throughout the world for at least 5,000 years.[37] Interesting historical uses for onions abound, including—get this—as muscle toner for Roman gladiators who reportedly rubbed their bodies with onion juices before fights.[38] If their appearance wasn't ferocious, I bet the smell was. To be clear, this section refers to common bulbing onions, *Allium cepa*—which includes shallots (*A. cepa* var. *aggregatum*)—grown to maturity. (Bunching onions are the same species, and sometimes the same varieties, but they're harvested before full maturity.)

Onions are staples of many food cultures, partly because they can be grown and stored in so many climates. In my home, we're often cutting and sautéing onions before we even know what we're cooking. Their ubiquity leads to customer demand, but I caution growers new to storage to get into onions slowly. Depending on your climate, the curing process can be tricky to get right, and cleaning onions after drying is no small task when done by hand. And unless your yields are stellar, onions are a difficult crop to make money with, especially since onions are so cheap on the commodity market. That all said, growing and storing successful onion crops is rewarding and fun, and customers appreciate the local option!

Variety Selection

Not all onions are equal in storage. Pungent varieties—those that make your eyes burn—have longer storage lives than sweet varieties. It's common for well-cured pungent onions to keep in excess of six months, whereas sweet varieties rarely last longer than three. There's also photoperiod to worry about. In the United States, seed and plant companies usually sell onions as short-, intermediate-, or long-day based on the day length required to initiate bulbing. Long-day onions generally store longer than short-day onions,

but your location limits what you can successfully grow. If you're south of a variety's acceptable latitude range, the onions may not form bulbs or reach dormancy. If you're north of a variety's latitude range, the bulbing process will begin too early, and your onions will be small.

Within each day-length category, different varieties take more or less time to mature. Because proper curing is vitally important to storing onions, it's a good idea to choose varieties that mature during weather conditions conducive for curing (warm and dry). Growing onions from sets rather than from seeds or plants can help far-northern growers shorten time to maturity to grow a mature onion crop with time left for curing.

Cold Hardiness

I'm rarely fighting frosts when onions are maturing, but I often leave onions in the unheated spaces used for curing until cold weather threatens to damage them. Like many other storage crops that are mostly aboveground, onions can handle light frosts but are easily frozen when exposed to prolonged below-freezing temperatures. Onions can begin to freeze below about 30°F (−1°C) and show damage as water-soaked tissue. Often, outer layers freeze first, but sometimes inner layers show damage before outer layers freeze.[39]

Harvest Indicators

Onions grown for storage go through three general growth phases: a vegetative phase when leaves and roots develop, a bulbing phase when bulbs form and expand, and a ripening/resting phase when onions suspend further growth and dry down as they enter a state of dormancy. As onions approach their resting phase, the necks slowly soften (they give when pinched) and the tops eventually fall over. Onion bulbs continue growing as the necks soften, albeit more slowly, until the tops fall over. After tops are down, the outermost layers (called scales) and the tops dry out because the plants no longer direct water toward those parts. In warm and dry weather, the tops and outer scales dry quickly, but in cool, damp weather, the process is slow and gives pathogens a ripe opportunity to enter the bulbs through the necks.

Timing the harvest is about balancing yields and storage quality, and most farmers make this judgement based on the percentage of fallen onion tops in their field. If you wait until all the tops fall, you maximize yields but risk more problems with disease and rot in storage. If you harvest too early, before many tops have fallen, you risk lowered yields and immature onions that are difficult to dry and continue growing

Photo courtesy of Phil Knapp.

in storage. If long-term storage is the goal, most sources agree that the best time to harvest is when 50 percent of tops have fallen, although some advice says to wait until 80 percent of tops have fallen to increase overall yields.[40] Delayed harvests, however, have been shown to increase levels of hormones that can cause earlier sprouting in storage.[41] That said, many farmers in warm and dry climates allow onions to fully dry in the field before harvesting.

Note that by harvesting, I mean pulling onions from the ground, but not necessarily removing them from the field. Field curing onions is common practice for many farmers (see below). It's also common to withhold water for one to two weeks before harvest to aid maturation and drying.

Trimming and Processing

As with garlic, most customers expect onions with tops removed. Many farmers harvest and cure onions as whole plants, tops and all, whereas others remove tops before curing in the field or indoors. In trials conducted in New Zealand, it didn't matter when onions were topped unless it rained during field curing, in which case leaving plants intact was better.[42] I remove onion tops in the field before curing indoors to leave behind some new organic matter and to reduce the volume I need to haul. I typically leave roots intact for the curing process because I feel they help remove moisture from the bulb.

Cleaning onions after curing can be a lengthy process, during which you remove any leftover tops above the necks, remove dirty outermost scales, and possibly clip the roots. Farmers who grow a lot of onions often use onion topping machines and brushing tables to clean their onions more quickly (see "Trimming," page 31).

Curing

In onions, curing dries the outermost scales and narrows the necks to create a protective barrier against water loss and pathogens. If the weather is dry and warm, onions may cure in the field after harvest until leaves, necks, and outer scales feel papery and dry. In dry weather, this may take one to two weeks. Wet weather will delay the drying process and increases the risks of developing fungal diseases in storage.[43] My rule of thumb is to bring onions inside for anything more than a passing shower, but some farmers are more tolerant of wet weather (see "West Farm," page 203).

If you're curing onions indoors (say, in a shed, greenhouse, garage, or similar structure), good conditions are between about 80 and 90°F (27 and 32°C) and around 65 percent RH. Use fans to force air through and around the onions. (In the rainy United Kingdom, it's common to cure onions at 82°F (28°C) and 65 percent RH for six weeks.[44] Throughout much of Scandinavia, it's common to cure onions at 68°F (20°C) with the goal of maintaining scale quality and color at lower temperatures.[45]) Many farmers set onions out on wire racks, but it's possible to cure onions in bulk bins up to 1,000 pounds (454 kg) with adequate airflow. Cure onions until their outer layers and necks are completely dry and papery. Depending on the temperature, humidity, and their dryness at the start, curing may take a week or a month. Exposing onions to temperatures above 104°F (40°C) for more than three or four days during curing might damage them.[46]

Storage Conditions

Onions respond best when kept in dry conditions, between about 60 and 70 percent RH. The best storage temperatures are near freezing at 32°F, but many farms keep onions at higher temperatures, from about 40 to 60°F (4 to 16°C). These higher temperatures shorten storage life and increase the risk of early sprouting, but many farmers still manage to keep storage varieties in these conditions for three to five months. In ideal conditions, pungent storage varieties can last anywhere from six to nine months.

You may also hear about farms storing onions at high temperatures, especially in hot climates. Temperatures above 86°F (30°C) inhibit both sprouting and rooting, but relative humidity in storage must be higher, between 80 and 90 percent, to prevent onions from drying out.[47] Farms that produce onion sets often store them in conditions like these to preserve the onions without vernalization.

Onions need very breathable containers in storage to allow moisture from respiration to escape. At smaller scales, farms often use plastic mesh or burlap sacks that are breathable and stackable. At larger scales, ventilated bulk bins are a common choice.

Ethylene

Onions don't produce much ethylene, but exposure to very high levels can induce sprouting.[48]

Common Problems

When onions sprout either roots or shoots, they quickly lose moisture, becoming soft and shriveled. Neck rot is a common problem and can indicate problems with your harvest timing and curing. Symptoms include bulbs that soften and turn brown from the top down. Onions left too long in the field, especially during wet weather, are susceptible to neck rot. The same goes for onions that weren't dried quickly enough. The fungus *Botrytis allii* is the most common culprit of this disease, and it spreads down the foliage or neck through wounds into the bulb. If you're topping onions before curing, keeping the remaining tops 6 inches (15.2 cm) or longer can prevent the fungus from growing into the bulb before the neck dries.[49]

Parsnips

As writers for the International Society for Horticultural Science thoughtfully put it, "If carrot is the Prince Charming of the root vegetables, then parsnip is surely Cinderella, unloved, ignored, and rejected."[50] Indeed, before the advent of the potato, parsnips were one of the primary starch crops in Europe, but now they're relegated to the status of specialty ingredient. Parsnips (*Pastinaca sativa*) are relatives of carrots in the parsley family, but they carry a flavor and texture that's all their own. I think they're useful crops for any storage farmer. At my farm, they're customer favorites, probably because the parsnips in supermarkets usually look bad and taste bland. They also grow into deeper soil strata than anything else on the farm, loosening deep layers and bringing up nutrients. Some people, including me, get painful blisters after coming into contact with the leaves. Cultivated parsnips and other closely related plants, such as cow parsnip (*Heracleum maximum*), produce compounds in their leaves called furanocoumarins that, when exposed to sunlight, react with oxygen and subsequently damage skin cells.[51] If you're new to growing parsnips, wear long sleeves and gloves when working in the leaves until you know how you react.

Variety Selection

Parsnip varieties are generally cold hardy and keep similarly in storage. That said, some varieties tend to grow straighter and are more resistant to branching than others. Oxidative browning in storage differs between varieties (see below), but seed companies rarely mention this trait in descriptions.

Cold Hardiness

Parsnips are the most cold-resistant vegetable that I know of. The part we eat is almost entirely belowground; with some varieties, even the crowns stay safely below the soil. Their freezing point is usually between 29 and 30°F (−2 and −1°C), but their roots can lightly freeze without any signs of damage when thawed.[52] (Parsnips do sustain damage when frozen solid.) Some springs, I find surprisingly good-looking parsnips missed during harvest that survived a subarctic winter. For most growers, frozen ground and snow cover are bigger concerns than damaged parsnips when racing the onset of winter.

Harvest Indicators

You can harvest and store parsnips at nearly any size; aside from the issue of drying out, small parsnips keep just as well as large parsnips. In areas with longer growing seasons, oversized parsnips are possible. When wider than about 3 inches (7.6 cm) at the crown, parsnip roots develop a spongy texture in places, and you should harvest your parsnips before this happens. (Oversized parsnips are common when you give them too much space during thinning.) If your parsnips are not oversized, you can harvest them any time before cold weather makes harvest impractical, and preferably after a frost. Exposure to at least one frost makes parsnips sweeter, as they convert some of their starches to sugars.

Trimming and Processing

Trim off parsnip leaves flush with the crown and without cutting into the root. You will inevitably leave some bits of petioles behind because parsnip crowns are often concave on top. If you choose to wash parsnips before storage, be sure they're dry on their surfaces before putting them into any semi-enclosed container. Before selling parsnips, it's common practice to trim off parts of the long, thin taproots.

Storage Conditions

Parsnips store best at 32°F and around 98 percent RH. If the storage room is drier than this, use semi-enclosed containers, such as perforated bin liners, to limit the parsnips' exposure to dry air. In good conditions, expect parsnips to last for four to five months before they begin sprouting roots and leaves.

Ethylene

Parsnips do not produce much ethylene on their own, but they're moderately sensitive to exposure and form a bitter taste when exposed to too much for too long.[53]

Common Problems

Like carrots, parsnips are very susceptible to drying in storage if the humidity is too low. Oxidation (a.k.a. browning) is another common issue without a good solution. As parsnips age in storage, compounds in the outermost tissue oxidize to create yellow and brown hues that stain the white roots.[54] Trials have shown that certain varieties are more susceptible to oxidation than others; White Spear is a variety that shows relatively little oxidation compared to other varieties.[55]

Black canker is a disease common to parsnips in North America, Europe, and Australia, caused by fungi in the genus *Itersonilia*.[56] The brown-to-purple lesions around the crowns will be obvious at harvest time, so don't bring affected roots into storage.

Friends of Offbeet Farm helping with a snowy parsnip harvest.

Potatoes

Considering that potatoes (*Solanum tuberosum*) are the fourth most widely grown crop in the world (behind wheat, rice, and corn), it's surprising that they entered the world stage just a few hundred years ago. South American peoples have been cultivating potatoes for more than 12,000 years, but potatoes didn't enter Europe until the 1560s and East Asia until the early 1600s.[57] Their ease of cultivation, versatility, high starch content, and (importantly) storability quickly made them important food sources. They've also proven adaptable. Depending on how you group them, there are between 188 and 219 species of wild potatoes in the genus *Solanum*

growing across parts of southern North America, Central America, and South America, and breeders regularly introduce traits from wild species into new potato varieties.[58]

Potatoes are now ingrained in our food cultures, and I think most customers expect diversified storage farms to grow them. I do not currently grow potatoes on my farm. Instead, I buy them from another local farm that specializes in potatoes to supply my CSA. One reason is that growing potatoes takes a lot of soil disturbance, and I fear erosion in my steeply sloped fields. More importantly, I don't have the equipment to harvest them efficiently. The great thing about

Potatoes cut and drying before planting at Root Cellar Farm. (Also, notice our rooster Brad Pitt photobombing this image.)

potatoes, though, is their ubiquity. If you can't grow them profitably at your scale, look around for other nearby farms that can.

Variety Selection

There may be hundreds of potato varieties to choose from depending on the sources available to you! The sheer number of options for shapes, colors, and textures, not to mention yields and disease resistance, can be daunting. While storage life is potentially another factor to consider, most potatoes store remarkably well when harvested healthy at full maturity and properly cured. There are some minor differences to be aware of, but these follow familiar patterns. Generally speaking, early-crop (early maturing) potato varieties don't store as long as late-maturing varieties. There are exceptions, and the storage lives partly depend on how long the potatoes remain dormant before sprouting. Disease resistance and resistance to injuries like skinning and bruising during processing also help prevent rots that shorten storage life.

Cold Hardiness

The aboveground foliage of potato plants is extremely frost sensitive. Any exposure to frost will kill surface tissue on leaves and stems and accelerate the dieback process. Potato tubers are also very sensitive to freezing, but they have the benefit of being underground. That said, it's wise to harvest your potatoes well before the ground starts freezing. Temperatures near 32°F may injure certain potato varieties within a few weeks, but exposure to frozen soil or below-freezing air temperatures (during harvest and processing) often damages potatoes both internally and externally.[59] Previously frozen tissue often appears water-soaked and sunken before turning brown, but freeze damage sometimes manifests as internal lesions that are impossible to see without cutting into the tubers.[60]

Harvest Indicators

One danger of harvesting potatoes too early is higher respiration rates, which lead to shortened storage lives. Another is immature skins, which promotes bruising and skinning injuries that can fester in storage. Farmers usually sell new potatoes—those harvested before full maturity—soon after harvesting, but it's possible to store them for about four months with careful handling and curing.[61]

Mature potato tubers have reached their full size and stopped growing, have converted most of their stored carbohydrates to starch, have begun the process of skin setting, and have entered a state of physiological dormancy.[62] All of these factors combine to make mature potatoes last longer in storage than new potatoes, but how can you tell if your potatoes are fully mature? Many farmers use the foliage and skins as indicators, as well as past experience. As potato foliage begins to die back naturally, tuber growth slows to a stop, and the potatoes enter a state of hormonal dormancy. The skins also begin setting as the aboveground plant dies. If the vines have died back naturally and without the onset of frost, the tubers are ready for harvest. If sudden frosts kill green, healthy vines, you should wait two to three weeks before harvesting. The skins should feel firmly attached and resist flaking off when you rub the tubers.

The modern potato industry often uses artificial vine killing as a strategy to make the tubers mature early, and you can use this strategy too. Killing potato vines causes the same physiological changes in the tubers as natural dieback. Mowing, chopping, pulling, and flaming all work. Depending on the weather conditions, you should wait two to three weeks before harvesting after killing vines.[63]

Trimming and Processing

Potatoes typically don't require any trimming. If washing potatoes before storage, it's important that they dry before or during storage to prevent fungal or bacterial problems (especially the anaerobic ones).

Curing

During curing, potatoes go through a process of setting their skins and healing wounds. The latter is commonly called suberization, wherein the tubers create waxy barriers over cell walls and a protective compound called suberin that impedes pathogens.[64] Overall, curing makes potatoes less susceptible to skinning injury, water loss, and disease from pathogens.

Most farmers cure potatoes where they're stored. Although potatoes can cure at temperatures anywhere between about 55 and 75°F (13 and 24°C), most sources agree on 59°F (15°C) as the optimal curing temperature because it minimizes decay while keeping the curing period relatively short—around one to two weeks. Humidity should be very high during curing, and anything below 80 percent RH will delay the process.[65] Potatoes need excellent airflow during the curing process. With small containers, farmers can pull air through the bins with fans, but in large bulk piles, perforated vent pipes might be needed to deliver fresh air to the center of the piles.

Storage Conditions

After curing, the temperature should be slowly ramped downward by about 0.5° F (0.3°C) per day to the target temperature. This slow ramping achieves two things: (1) It prevents condensation from forming on the potatoes, and (2) It reduces the incidence of pressure bruising in large piles or containers. For most of us small-to-medium-scale farmers selling potatoes for regular consumption and occasionally seed, the optimal storage conditions are between 38 and 45°F (3 and 7°C) with over 95 percent RH.[66] Storing potatoes closer to 45°F will limit the development of sugars that lead to browning during frying, while storing potatoes closer to 38°F will improve storage life by minimizing respiration rate. I'll let you decide. You may read about industrial potato storage above 45°F. Potato processors making potato chips and french fries prefer these storage temperatures to minimize sugars that cause browning during frying, but they often treat potatoes with sprout inhibitors to make storage at these warmer temperatures possible.[67]

You can store potatoes in any containers that allow excellent airflow, such as bulb crates or plastic macrobins. It's also common to store potatoes in bulk piles, but you'll need a means of ventilating through the piles, such as perforated pipes or slatted floors.

Ethylene

Intact stored potatoes produce very little ethylene, but injuries increase the amount produced. Potatoes in storage have complicated responses to ethylene exposure. Upon initial exposure, potato tubers tend to ramp up their respiration and may lose moisture, but respiration declines with duration of exposure and at lower temperatures.[68] Ethylene also appears to break tuber dormancy and induce sprouting, but continued ethylene exposure slows the growth of sprouts from eyes, leading to short and thick, slow-growing sprouts.[69]

Common Problems

While potatoes often perform great in storage, there are a lot of potential diseases that cause storage rots, including pink rot, late blight, dry rot, soft rot, silver scurf, black dot, and early blight.[70] Chilling or freeze damage can cause similar symptoms, with softened discolored flesh that deteriorates quickly. The problem of secondary rots—when the decay of one piece leads to the decay of those around it—is not unique to potatoes; I've seen it happen many times in stored carrots, beets, and cabbages, to name a few. With potatoes, however, primary and secondary rots are particularly gross. I can only describe the odor of rotting potatoes as the smell of death. The smell, however disgusting, can be helpful in notifying you of a problem you can't yet see.

In large containers or deep piles, bruising or flattening caused by the pressure of the potatoes above can lead to losses. Properly curing your potatoes, maintaining high humidity in storage, and ramping to storage temperatures slowly can all reduce the severity of pressure bruising.

Rutabagas

Rutabagas—also known as swedes—are often confused with turnips, but they're an entirely different species. They're members of the same species, *Brassica napus*, as canola and Siberian kale. I had long heard that rutabagas originally came from a cross between turnips (*B. rapa*) and cabbage (*B. oleracea*), but apparently not. New research from the University of Missouri points to rutabagas sharing an ancestor with, rather than descending from, turnips and cabbages.[71] People first widely cultivated rutabagas in Swedish-controlled Finland in the 1500s, and rutabagas became associated with the Swedish people as the vegetable spread, hence the common name swede.[72] The more common name, rutabaga, comes from a local dialect spoken in southwestern Sweden, comparing the large, awkwardly shaped roots to rams.[73] (The Swedes made a similar comparison between male sheep and Norwegians.)

Depending on where you live, your customers might need some education on rutabagas and their uses in the kitchen. When I lived in Michigan's Upper Peninsula, rutabagas were well known and popular as essential ingredients in pasties. My Alaskan customers were less familiar with rutabagas but almost universally learned to adore their sweet and distinctive flavor. Although similar to turnips, rutabagas have stronger flavors, are more resistant to cold temperatures, and last longer in storage.

Photo courtesy of Phil Knapp.

Variety Selection

To my knowledge, different varieties of rutabagas store similarly. Where they differ is their shape, internal coloring, and disease resistance. Depending on your location and disease pressure, clubroot might make growing nonresistant varieties impossible. For example, the widely popular variety Laurentian is not resistant to clubroot, whereas the relatively obscure variety Marian is.[74]

Cold Hardiness

Similar to other root vegetables with large portions that stick up out of the ground, rutabagas are damaged by sustained temps below their freezing point, between 29 and 30°F (−2 and −1°C).[75] That said, rutabagas grow tall and expansive foliage—in fact, they're commonly used as a grazing crop for cattle—that can protect the roots from frost, and rutabagas are hardier against cold than similar veggies like beets, turnips, and radishes. Harvest the more sensitive crops first if you're forced to triage.

Harvest Indicators

You can harvest rutabagas for storage at nearly any size, but customers prefer large and round roots. Waiting to harvest until after a frost will improve their flavor.[76]

Trimming and Processing

Whereas most other root veggies grow their leaves from very abbreviated stems that are indistinguishable from the tops of the root crowns, rutabagas sometimes, but not always, grow a slightly elongated stem. When these appear, I usually make my cuts somewhere between the lowest attached leaf and the root crown. Rutabagas also tend to carry soil in bunches of fine roots near the bottom of their enlarged hypocotyls (in other words, the part we eat). I usually trim these off too, cutting slightly into the flesh, to avoid transporting so much soil and to make washing easier. I also trim off mild tunneling from root maggots when present. So long as the cut surfaces are small, the rutabagas don't suffer from the trimming in storage. As with other root veggies, make sure they have an opportunity to dry if you wash them before storage.

Storage Conditions

Rutabagas keep best when the temperature is 32°F and RH is at or above 98 percent. Containers with good airflow are best, as this allows leftover leaf bases to dry rather than rot as they senesce in storage. Rutabagas are not as susceptible to drying out as carrots, parsnips, beets, and turnips, but will eventually soften if humidity is too low. In good conditions, expect rutabagas to last for four to six months before rotting or sprouting new leaves.

Common Problems

Brown rot (also known as *Phytophthora* storage or root rot) can manifest in stored rutabagas grown in fields where other brassicas have problems with fungi in the genus *Phytophthora*.[77] The first signs of decay usually appear at the trimmed stems and eventually work downward into the roots. Early decay can come from leaf bases leftover from trimming (similar to kohlrabi). Watch out for unusually lightweight roots and those with rough-textured skin, as these are usually stringy and bitter or are hollow within.

Sunchokes

Sunchokes (a.k.a. Jerusalem artichokes) are the species *Helianthus tuberosus* and members of the Aster family. They are close relatives to common sunflowers (*H. annuus*) but likely originated from several species of perennial sunflowers found across eastern North America.[78] Several Native American groups, including the Cree and Huron, grew and bred sunchokes for their tubers, which vaguely resemble ginger roots.[79] I choose not to call them Jerusalem artichokes because the name is misguided both geographically and botanically. As the story goes, the explorer Samuel de Champlain thought the tubers tasted similar to artichokes—which I question—and the moniker unfortunately stuck. After their introduction to Europe, the Italians named them *girasole articiocco* (sunflower artichokes), and the English corrupted that name to Jerusalem artichokes. I prefer sunchoke because it nods to the plant's origin and its membership in the sunflower genus, *Helianthus*.

Although we harvest, store, and eat the tubers, the aboveground plants can be terrifically productive, forming tall hedges with bright yellow flowers. I fell in love with sunchokes and their unique, nutty flavor while living in Sweden, where you can find them for sale in most groceries. They are disappointingly difficult to find in the United States despite originating here, and you might have to educate your customers to get them onboard.

Variety Selection

Honestly, there are not many variety options out there for sunchokes. For better or worse, most of the breeding efforts in the last half-century have focused on inulin, an undigestible but sweet carbohydrate found in sunchokes and used in artificial sweeteners and prebiotics. There are a handful of culinary varieties that have traits like more uniformly large and round tubers and tubers that grow compactly under the plant rather

than ranging several feet through the soil. You'll be hard-pressed to find seed companies that offer more than one type, and variety names are often omitted. There are red-hued variants out there, but I cannot speak to their storability as compared with the white types. My advice is to get your hands on whatever sunchokes you can! Pay attention to tuber shape and size. Sunchoke tubers can be knobby, and the easiest ones to process (and cook) are more uniformly round in shape. If you're in the far north, also pay attention to photoperiods. Most sunchokes do not begin setting flowers and tubers until daylength drops below 13.5 hours.[80] Stampede is a day-neutral variety that will begin setting flowers and tubers earlier than other cultivars.

Cold Hardiness

Sunchokes are another supremely cold-hardy vegetable. In fact, they're perennials in many locations, though we typically grow them as annuals. Most sources list them as hardy down to USDA hardiness zone 4, but their natural range takes them into zone 3 and potentially lower. When I grew them in the Upper Peninsula of Michigan (zone 4), volunteers came back every year. I haven't tried them at my zone 2 farm near Fairbanks. I group them with parsnips as a vegetable you can dig out of frozen ground without worrying about freeze damage. In fact, the USDA suggests leaving them in the ground all winter if frozen soil won't prevent you from harvesting them as needed.[81]

Harvest Indicators

As noted above, many sunchoke varieties don't begin setting tubers until daylength drops below 13.5 hours. When I grew them, I left them in the ground for as long as possible while harvesting remained practical. Like other sunflowers, frost causes the aboveground plants to begin dying back, and it causes physiological changes in the tubers that make them sweeter. You can harvest them any time after flowers appear, but it's best to wait until the foliage dies back and preferably longer.

Sunchoke tubers are typically smaller than potatoes. Some folks suggest modifying potato harvesters to accommodate the smaller size—for example, by adding additional metal fingers to shaker-style diggers. If digging by hand, running an undercutter first will be helpful. The tubers often, but not always, congregate under the plants. Some wild types, however, grow tubers several feet away from the parent plants.

Trimming and Processing

Depending on when you harvest, you may need to trim the stolons/rhizomes connecting the tubers to the plant. If left in the ground, these will eventually senesce and fall off. You can store sunchokes either washed or dirty, but you should allow them to dry before storage if you wash them first. The tubers are not particularly sensitive to damage and can easily tolerate tumbling in a barrel washer.

Storage Conditions

Sunchokes store best when kept at 32°F with an RH near 95 percent. In these conditions, sunchokes should last for three to four months before losses exceed 25 percent.[82]

Common Problems

Sunchokes are very susceptible to both drying and pathogens in storage. Keeping the temperature as close to 32°F as possible will limit pathogen grown. Keeping humidity above 90 percent will help prevent moisture loss, which accelerates respiration and further deterioration.[83]

Other Considerations

Sunchokes are aggressive plants and will readily become a weed if not controlled. At West Farm in Vermont, areas planted to sunchokes are either left fallow or aggressively cover-cropped the following year to control volunteers (see "West Farm," page 203).

Sweet Potatoes

Sweet potatoes (*Ipomoea batatas*) are members of the morning glory family (Convolvulaceae) that were first cultivated in northern South America nearly 5,000 years ago.[84] Today they're one of the most widely grown crops in the world and are generally adored by customers. If you can successfully grow sweet potatoes, especially in the north, there will be a market for them. Sweet potatoes are well-known warm-weather crops. Before interviewing other farmers for this book, I didn't think of sweet potatoes as something northern farmers could grow (see "West Farm," page 203). Now they're something I'm even thinking of trialing in interior Alaska, although I'll likely need some forms of season extension to succeed.

Sweet potatoes are linked to some interesting history. There's compelling evidence that they spread to South Pacific islands well before the Spanish brought them to Europe. Archaeologists suspect that Polynesian sailors acquired sweet potatoes in South America sometime in the thirteenth century and spread them to Easter Island, Hawaii, Polynesia, and as far west as New Zealand.[85] (This explains why peoples in South America and the South Pacific islands have similar words for sweet potatoes in their native languages: *cumal* and *kumara* in northwestern South America, *'uala* in Hawaii, *'umala* in Samoa, and *kūmara* in New Zealand, to name a few.[86])

Today in the United States it's common to see sweet potatoes sold as yams. Botanically, sweet potatoes and yams are very different. Yams belong to an entirely different family (Dioscoreaceae) and are more closely related to onions and palm trees than they are to sweet potatoes. The starchy tubers originated in Africa and are commonly grown in the tropics. So why do we call sweet potatoes yams? The word *yam* is close to words in several West African languages that mean "to eat" or, in one case, a word that refers to the starchy tuber itself. Sweet potatoes were staples in the diets of many enslaved Africans after they were taken to the American South. Through generations of the enslaved, the word yam slowly shifted from its original meanings to refer instead to sweet potatoes. Much later, in the 1930s, marketers were looking to rebrand new varieties of sweet potatoes to the American public, and they chose *yam* as something already familiar to many in the American South.[87] Nowadays, the word yam is synonymous with sugary orange sweet potatoes. If you look at a can of yams from a US supermarket, the first ingredient will be sweet potatoes.

Variety Selection

Most sweet potato varieties do well in storage. North Carolina State University describes Averre, a variety the university was involved in developing, as not well suited for long-term storage. The North Carolina Crop Improvement Association writes that Averre "should only be considered a 6–8 month storage cultivar."[88] That might seem like a long storage life, but some sweet potato cultivars will store up to one year when properly cured and kept in the right conditions.[89] In particular, trials have shown that white varieties are the slowest to lose moisture in storage, whereas orange

Photo courtesy of Jillian Mickens.

varieties are uniformly sweeter than white varieties.[90] For northern growers, producing a crop is more important than the storage potential of a given variety. A trial by Pennsylvania State University found that marketable yields nearly doubled when they let sweet potatoes grow for 120 days rather than 90 days before harvest, and the fastest-growing varieties—Beauregard, Averre, and Orleans in their trial—gave the best yields regardless of the harvest date.[91]

Cold Hardiness

Sweet potatoes are very intolerant of cold both in the field and in storage. Sweet potato vines are completely intolerant of frost, but since we're after the roots, a light frost is usually okay. Sweet potato roots sustain chilling injury at temperatures lower than 54°F (12°C). Chilling injury is cumulative, and sweet potatoes are more susceptible before they're cured. One trial found that after two weeks at 45°F (7°C), about 10 percent of noncured sweet potatoes were injured, and after four weeks nearly all the roots showed signs of injury.[92] Injured roots usually show surface pitting, shriveling, and fungal decay.

Harvest Indicators

Sweet potato roots will keep growing indefinitely—or until they're so large that they rot from the inside—so long as there's no frost and growing conditions are good.[93] For short-season growers, oversized roots are rarely a problem, and either frost or low soil temperatures will force you to harvest instead. Frost kills the aboveground vines, and decay may spread from the vines into the roots if left unharvested for too long. In the absence of frost, you should harvest the roots before soil temperature drops below 54°F to avoid any chilling injuries.

Trimming and Processing

Sweet potatoes themselves do not need trimming, but it's common to cut back, mow, or remove the vines prior to harvest to make digging easier. The roots are sensitive and easy to damage between harvest and storage. Many farmers forgo barrel washers and use spray tables or rinse conveyors instead.

Curing

Like potatoes, sweet potatoes go through a process of skin setting and wound healing (suberization) during curing.[94] While curing, sweet potatoes also rapidly turn their starches into sugars like sucrose, glucose, and fructose that persist throughout storage.[95] (They wouldn't taste very sweet without the curing step.) Unlike regular potatoes, sweet potatoes cure at higher temperatures and for less time. Curing has been most successful at temperatures between 82 and 90°F (28 and 32°C) and at or above 90 percent RH for four to ten days.[96] Airflow is very important during the curing process to ensure roots have enough oxygen for respiration.

Storage Conditions

The best conditions for storing sweet potatoes are temperatures between 57 and 60°F (14 and 16°C) with humidity between 85 and 90 percent.[97] Despite the danger of chilling injury below 54°F (12°C), don't be tempted to keep them at warmer temperatures, because temps above 66°F (19°C) will induce sprouting within a couple months. Sweet potatoes need containers with good ventilation because of the risk of mold at humidities near saturation. Despite their sensitivity, it's okay to store sweet potatoes in bulk containers such as macrobins, so long as you load them carefully.[98]

Ethylene

Sweet potatoes generally don't produce enough ethylene during storage to damage themselves. (And they're usually stored alone.) If, for whatever reason, sweet potatoes are exposed to high levels of ethylene, they typically lose some sweetness and become discolored.[99]

Common Problems

The majority of decays and rots in stored sweet potatoes result from infected injuries incurred during harvest, washing, and processing. It's important to treat the roots gently, remove damaged roots, and cure them properly. Some growers sanitize their sweet potatoes prior to storage to limit the number of surface pathogens brought inside.

Turnips

Turnips are the same species, *Brassica rapa*, as Napa cabbages and bok choi, but something close to the turnip form was the first variant developed from wild species. Turnips as root crops date back to at least 1800 BCE, and likely earlier.[100] Throughout much of recent history, turnips have gotten a bad rap. The phrase "turnip eater" was used in medieval Europe to describe someone as a country bumpkin, rube,

clodhopper, hayseed, and the like, and Charles Dickens used turnip as an insult similar to "idiot."[101] Turnips were food for the masses and were largely replaced when potatoes spread across Europe and Asia. But here's the thing: I like turnips. I'm eating one as I write—one perk of being a storage farmer. I wish kohlrabis had the texture and flavor of turnips (apologies to kohlrabi-lovers).

Turnips may not be the most popular vegetable out there, but I encourage storage farmers to grow and promote them to customers. They have vibrant colors, are generally productive, and perform admirably in storage. People usually don't eat many turnips because they're not large parts of our food cultures, but I've found that customers respond enthusiastically to turnips when given the chance.

Variety Selection

When selecting turnips for storage, it's important to differentiate between salad turnips and storage turnips, although the distinctions aren't always clear. I use a variety's maturation time as an approximate guide: The more days to maturity, the longer it will last in storage. If you're struggling to choose, stick with classic storage varieties such as Purple Top or Golden Globe.

Cold Hardiness

Turnips behave similarly to other root crops with large parts of their swollen hypocotyl sticking up and unprotected above the soil surface. While turnips can handle several rounds of light frosts, sustained exposure to temperatures below their freezing point of 30°F (−1°C) will cause irreversible damage to the turnip, showing as water-soaked tissue with a blistered appearance on the skin.[102] Some varieties have lush greens that might protect the roots below from light frost events.

Harvest Indicators

You can harvest turnips for storage at any size, but customers generally prefer turnips larger than golf balls, or about 1.5 inches (3.8 cm). Some turnip varieties tend to turn pithy or hollow when overlarge.

Trimming and Processing

Trim turnips in a similar manner to radishes and beets, removing the leaves as close as possible to the root crown without cutting into the flesh beneath. Turnips usually don't carry as much soil with them as rutabagas, but you can trim off the root tip prior to storage if you want that aesthetic. Damage from cabbage root maggots can be problematic in areas with the pest. You can trim off damaged portions of the roots, but these cut surfaces tend to lose moisture and oxidize, turning brown or grey, within a month or two in storage. If washing turnips before storage, be sure they can dry either before or during long-term storage.

Storage Conditions

Turnips keep best in cold, damp conditions, near 32°F and above 95 percent RH. Turnips are not as susceptible to drying as carrots and parsnips, but they will soften if the humidity is too low. Containers that allow some airflow will prevent excess moisture from forming on and around the turnips that can harbor mold growth. In good conditions, turnips should last at least four to five months in storage.[103]

Common Problems

Turnips suffer from similar pathogens to rutabagas, including *Phytophthora* storage rot. Fungal infections are common, especially by fungi from the genera *Phoma* and *Sclerotinia*.[104] Turnips are less likely than rutabagas to develop stringiness in the root.

Winter Squash and Pumpkins

I remember one year when an irrigation problem led to a near-total crop failure in my winter squash and pumpkins. Apart from my obvious distress at losing the crop, I felt sad without the colorful fruits adorning my field, storage room, and dinner table that fall and winter. To me, ripe squashes and pumpkins are quintessential to the fall experience. I know I'm not alone in these feelings. Maybe we can blame Starbucks and their pumpkin-spice lattés, but many customers expect farms to grow these colorful cucurbits each autumn. I'm also well aware that the economics of squashes and pumpkins, especially on small farms, can be dismal. On an area basis, winter squash and pumpkins can be some of the least profitable crops. That's further exacerbated by their volatility in storage. Despite this, I continue to grow them, mainly because they bring joy to both me and my customers.

Winter squash and pumpkins originate from species in the genus *Cucurbita* native to Central America, and they have been cultivated for nearly 8,000 years.[105] Similar to sweet potatoes, with origins in tropical and subtropical regions, squash and pumpkins prefer warmer growing and storage conditions than most other storage crops and are injured by low above-freezing temperatures.

Variety Selection

Some of my favorite pages in every seed catalog are those for winter squash and pumpkins. The diversity always amazes me. In fact, I've read that one species of squash and pumpkins, *Cucurbita pepo*, has the most diverse fruit morphology of any plant species on Earth. Aside from mitigating risk, growing a diverse array of squash and pumpkins can fill niches in the storage season. Your local climate, soils, and pests will help you decide what you can grow and store successfully. Your tastes also matter, as do those of your customers.

There are five domesticated species of *Cucurbita*, and of those, three produce storage varieties that most people will recognize. Each of those species have subtle differences in the way they grow and store. By growing different species and varieties, you should be able to supply customers with changing options through an approximate six-month storage period.

The most iconic and recognizable pumpkins and squash are in the species *C. pepo*, which originated in Mexico. This species includes, in addition to the most common types of summer squashes, most field and pie pumpkins, and common winter squash types including delicatas, acorns, and spaghetti squashes. Compared to the other species, fruits in *C. pepo* tend to be small and don't last as long in storage. Pie pumpkins, acorn squash, delicatas, and spaghetti squash usually don't last longer than three months after they reach maturity. The acorn subgroup should not be cured at high temperatures like other squashes; they tend to lose their dark green color in favor of yellow and develop off flavors and textures.

Butternuts, cheese pumpkins, and the strangely bumpy Futsu squash are members of the species *C. moschata*. Most customers recognize neck types like butternuts, which have lots of flesh and relatively small seed cavities. Some varieties, especially the butternuts, can store up to six months past maturity. Far-northern growers (those nearing latitude 60° north and above) tend to have trouble growing *C. moschata*, especially butternuts. This might have something to do with photoperiods affecting fruit-set.[106]

The final common species of winter squash and pumpkins is *C. maxima*. As the name suggests, these can grow to enormous sizes. The world's largest pumpkins—the current world record is 2,749 pounds (1,247 kg)—are varieties of *C. maxima*.[107] Of course, not all types of this species are so gargantuan. Common types include kabochas, buttercups, bananas, Hubbards, turbans, and, of course, mammoth pumpkins. Many of these are on the larger side for edible squash, and many have long storage potentials of six months or more. These squashes often improve their flavor in storage as they continue converting starches to sugars.

Cold Hardiness

Although some winter squash and pumpkin fruits can handle very light frosts, I think exposing them to frost is risky business. If your squash are mature, have relatively hard rinds (as with butternut squash), and there's excellent leaf coverage, you can probably get away with it. The foliage will die after any frost exposure, and it's easier to find and harvest squash after the foliage dies back some. That said, I don't ever risk exposing my squash and pumpkins to frost. Frost damage, when it appears, is disastrous for storage life.

In addition to frost, squash and pumpkins are sensitive to chilling injury below about 50°F (10°C). For this reason, I try to harvest my winter squash before daily mean temperatures drop below 50°F. Chilling injury is cumulative and commonly manifests as pitting and discoloration that eventually harbors infections such as *Alternaria* rot.[108] While any exposure to low temperatures results in small amounts of damage, the extent of chilling injuries depends on both temperature and duration of exposure. Visible damage often appears after three to four weeks at 40°F (4°C) and six to eight weeks at 45°F (7°C).[109]

Harvest Indicators

Winter squash and pumpkins are nonclimacteric fruits, meaning they stop maturing when removed from the vine. As such, immature fruits have no hope

Photo courtesy of Phil Knapp.

of developing the necessary changes for long-term storage once harvested. Squash and pumpkin fruits are generally mature and ready for harvest soon after they stop expanding (growing in length or diameter).[110] Some of the first visible signs of maturity are changes to the fruit rinds. When maturing, the rinds often change from shiny to dull as protective waxes form. This change often coincides with color changes and the development of a ground spot in some varieties. You'll also notice that small hairs, or trichomes, disappear as the fruits mature.[111] In my short Alaskan growing seasons, I'm often harvesting squash in this stage of development when cold temperatures force me to harvest.

The next visible changes are hardening of the rind, drying/corking in the peduncle (stem), and dieback in the foliage. These signs are less ambiguous than the waxiness and hairiness of the rind, which can be difficult to judge. If your squash don't have corky stems or tough rinds at harvest time, these changes can happen during curing. Fruits also develop the carotenoids that give the flesh deep orange and yellow coloring as they reach maturity; if you're willing to sacrifice a few fruits, the color of the flesh is a good indicator of when squash are ready for harvest. Even if your weather conditions allow it, leaving mature squash and pumpkins unharvested may be a bad idea. One trial found that squash left on the vines two to four weeks past physiological maturity had a much higher incidence of storage rots compared to squash harvested just as they matured.[112]

Trimming and Processing

The only trimming needed with squash and pumpkins coincides with cutting them from the vine. It's important to retain a bit of stem, or peduncle, to protect the fruits from pathogens. The amount you leave can be aesthetic and practical. I prefer to leave short stems of an inch (2.5 cm) or less to minimize damage to other fruits during harvest and transport. If you're storing squash and pumpkins in bins or piles, this is especially important.

You may hear about people washing or sanitizing squash and pumpkins with bleach solutions or other sanitizers. Of the farmers profiled in this book, only one is currently sanitizing their squash, and they're using a rinse conveyor. Most think that the extra handling and moisture introduce too many additional chances for damage and subsequent storage rots.[113] I don't attempt to clean my squash and pumpkins before storage and use a soft brush to remove any soil before sale. (There's not much because I grow squash on plastic.) Trials using sanitizing hot water rinses for the purpose of reducing storage rots often show no positive effects.[114]

Curing

Most winter squash and pumpkin varieties need a form of curing before storage. The process heals wounds, toughens the rinds, converts starches to sugars in the flesh, and, if necessary, pulls moisture from the stems. Collectively, these changes improve storability by reducing moisture losses and risks of infection from external pathogens. General advice is to expose squash to temperatures between 75 and 81°F (24 and 27°C) for one to three weeks after harvest.[115]

With all this in mind, know that curing winter squash and pumpkins is not black and white. Many farmers skip the curing step entirely because their climate allows for it. In places where the growing season is long enough for squashes and pumpkins to fully mature and fall weather is generally warm and dry, an additional curing step might not be necessary. Think stereotypical, glorious New England fall, with crisp days of blue sky and sunshine. In one famous trial conducted at Cornell University in central New York—this trial is heavily cited by the USDA and, by extension, most university extension services—researchers found no differences in the storability of squash intentionally cured and those brought straight into storage.[116] (The exception was a variety of acorn squash, which suffered from curing.) The farmers I interviewed in Vermont and North Carolina do not intentionally cure their winter squash. Likewise, universities conducting trials in eastern Oregon and central New York did not cure their squash either.[117] In these locations, I suspect the relatively long seasons and warm fall weather enable the physiological changes of curing to happen in the field.

In contrast, farmers with short growing seasons or cool, damp fall weather have to intentionally, and sometimes artificially, cure their squash and pumpkins. In western Washington state, Boldly Grown Farm is able to field-cure their squash in windrows. In northern Minnesota, Manitoba, and the Upper Peninsula of Michigan, the farmers I interviewed either cure squash in greenhouses or in heated sheds. In interior Alaska, I follow the advice of a long-time farmer and university extension agent who farmed in both Fairbanks and Nome, Alaska. He advises curing squash in a well-ventilated space at 70 to 80°F (21 to 27°C) for two to three weeks, saying, "Even for market sales, I do not bring squash to the market unless [they] have been cured first."[118] In addition to heating, I dehumidify the space to about 50 percent RH during the curing period. If I don't, I've found that stems do not adequately dry and storage rots quickly spread downward through the moist stems. If stems are already dry, curing at higher humidities—up to 95 percent—might reduce moisture loss in storage over a period of months.[119]

I advise anyone who feels doubt about curing to do it. With the exception of acorn squash, which shouldn't be cured, cured squash don't perform any worse in storage than uncured squash. In some cases, cured squash perform much better. Losing most of your squash crop in storage is a heartbreaking experience, and unless you're sure you don't need it, the extra effort of curing might spare you from that tragedy.

Storage Conditions

The best storage temperature for winter squash and pumpkins is between 50 and 55°F (10 and 13°C). Below that, they accumulate chilling injury, whereas higher temperatures speed up their degradation and cause green varieties to turn yellow.[120] Relative humidity in storage should be between 50 and 70 percent. At the drier end of the range, you'll see more moisture loss, whereas at higher humidities you'll see more losses from storage rots.[121] Storage life varies by species and variety but is generally in the range of two to six months. Many seed companies include expected storage life in their variety descriptions.

Winter squash and pumpkins need excellent airflow in storage to remove humid air that, when left stagnant, can accelerate storage rots. Farmers usually store squash either on well-ventilated racks in single layers or piled in vented storage bins. Given the relatively high incidence of storage rots, storing in a single layer allows you to find spoiled fruits sooner and reduces secondary infections. Using bins, on the other hand, saves space and handling time, but if a squash in the center of the bin spoils, you're likely to lose several pieces around it.

Ethylene

Winter squash and pumpkins don't produce much ethylene except when wounded. Exposure to high levels of ethylene tends to turn green varieties to a yellow hue, similar to storage at high temperatures.[122]

Common Problems

Immature squash and pumpkins will quickly lose moisture in storage, softening and developing a sunken appearance. Most decay in storage is caused by various fungi.[123] Wounds, inadequately dried stems, chilling injuries, and blossom scars are all common points of entry. Keeping humidity low will reduce the incidence of rots. Keeping storage temperatures steady will also prevent condensation from forming, which can readily trigger fungal infections.

In all the places I've farmed, I've fought against short and cool growing seasons to get fully matured squash. As such, my most common issues in storage have been with immature fruits and stem rots. I use the high-temperature curing period partly to weed out the immature and infected squash. It's also when I'm able to fully dry the peduncles; if they're not bone-dry when entering the regular storage period, I've seen huge losses from top-down rots.

Part 4

Storage Farm Profiles

Early in my farming career, a friend turned me on to the Farmer-to-Farmer Podcast hosted by Chris Blanchard. If it's unfamiliar to you, Chris Blanchard started a podcast as part of his agricultural consulting and education business, the Flying Rutabaga, after growing vegetables in Iowa for more than fifteen years. He interviewed over 150 farmers focused on organic and sustainable farming—the final episode count was 176—about all aspects of their farming and business practices and lifestyles. For the listener, each episode is like a mini farm visit. You hear farmers' stories and often dive into the minute details and challenges of their farms. The conversations are both inspiring and enlightening to budding and veteran farmers alike. It's hard to say just how many years' worth of mistakes I avoided because of the podcast; it undoubtedly shaped my development as a farmer. Sadly, Chris passed away from cancer in 2018. His legacy goes far beyond his innovative podcast; he showed a generation of farmers the value of connecting with each other, sharing knowledge, learning collectively from one another's mistakes, and working in community.

As I contemplated writing this book, I was thinking about Chris and ways to bring farmers together to share knowledge and experiences. Each episode of his podcast captured the essence of a farm visit, and I set out to accomplish the same thing in print. What better way to inspire people and show them what's possible with storage? Storage farmers from across the United States and one from Canada generously gave their time to describe their own journeys into winter storage, their successes and challenges, and to generally talk shop. I wrote up seven of these conversations into the profiles that follow, in addition to two profiles describing my past and current farms. The farms are in order of size; specifically, the poundage each farm stores annually.

For these profiles, I put myself into the boots of a brand-new farmer or someone new to winter storage. Honestly, this wasn't difficult; it hasn't been that long since I *was* a new farmer, and those anxious memories come back all too readily. What would I want to see and what would I ask about on a farm visit? It's these details I try to convey: Each farmer's motivations for getting into winter storage and why they continue with it; how and why farmers built their particular storage spaces; the fine details of harvesting, processing, and putting crops into storage; their marketing strategies during winter; and the thinking behind their decision-making. The profiles show what's working on real farms. The farmers were honest in sharing not only their successes, but some failures too. Everyone makes mistakes; hopefully you can avoid these particular ones while also treating yourself graciously when blunders inevitably happen. Each profile ends with some words of advice from the veteran storage farmers. I encourage you to take these to heart—they may be the most valuable parts of this book.

Root Cellar Farm

Owner: Sam Knapp

Location: Toivola, Michigan; about 20 miles south of Houghton in the Upper Peninsula

Farm size: ¼ acre devoted mostly to storage crops

Field setup: Permanent raised beds with 42-inch (107 cm) beds and 18-inch (46 cm) walking paths

Amount stored annually: 3,000 to 5,000 pounds (1,361 to 2,268 kg)

Signature storage crops: Carrots, onions, potatoes, squash, beets, and cabbages. Other crops include garlic, celeriac, rutabagas, and sunchokes.

Winter markets: Winter-only CSA and occasional wholesale deliveries

Summer markets: Occasional farmers markets and wholesale deliveries

Portion of annual sales from storage: Roughly 90 percent

Number of people working: Sam part-time

Root Cellar Farm was my first farm and first foray into growing storage crops for wintertime sales. In some ways, Root Cellar Farm was like an incubator farm. The land rental was effectively included with the house rental, and the owners did the initial plowing for me. Equipment costs were minimal: I rented a BCS tractor with a tiller and furrower for a weekend

Sam's mother Cindy walks in the Root Cellar Farm field. *Photo courtesy of Phil Knapp.*

to build the permanent beds, but afterwards used only a wheel hoe for cultivation. I also worked full-time as a graduate student and wasn't reliant on income from the farm. The farm's small size suited my needs while providing a low-stakes learning environment.

Root Cellar Farm demonstrates that profitable winter storage operations are possible even at very small scales. Relatively small harvests can fit into an unused corner or repurposed structure, and it takes minimal investment to get started. I found that testing the waters of winter storage at a small scale was a low-risk opportunity to learn techniques and trial the lifestyle changes that storage farming brings.

Storage Spaces

I was able to retrofit areas of a basement and stairwell for winter storage and processing. Our rented house had a large basement with a concrete slab floor and tall, 9-foot (2.7 m) cinderblock walls. The backside of the house had an auxiliary stairwell that connected the basement to the outdoors; this space became the root cellar. The entire length of the stairwell extended out beyond the side of the house and foundation. It was roughly 4 feet (1.2 m) wide by 12 feet (3.7 m) long with wide doors to both the outside and the basement. From the concrete stairs to just above ground level, the walls were cinderblock, and above that, uninsulated framing. A small gabled roof covered the structure and the stairs beneath.

I designed and finished the root cellar in the winter before I started production. Retrofitting this space to store winter vegetables was relatively easy and inexpensive. I insulated the stud cavities and the spaces between the ceiling joists with 4 to 5 inches (10–13 cm) of used EPS foam board and filled gaps with spray foam. I put a polyethylene vapor barrier over that, followed by used galvanized and corrugated roofing panels. I thickened both the basement and exterior doors with EPS foam insulation and used weather stripping to make airtight seals. I also shortened the exterior door (which wasn't used

much) to accommodate a framed space for a window air conditioner.

The stairwell initially lacked large, flat surfaces for stacking bins and boxes. L-shaped wooden stools—for lack of a better term—served as platforms for vegetable bins to sit over two stair steps. This way, I could fit four stacks of bins, holding roughly 80 pounds (36 kg) each, along one wall of the stairwell. I also built a platform measuring roughly 4 by 5 feet (1.2 by 1.5 m) in the empty headspace directly above the basement door. The platform was accessible by a removable ramp at the top of the stairs and allowed me to stack boxes in what had been wasted space. The total cost of these improvements was about $500 (in 2017).

To maintain optimum storage temperatures, the root cellar needed refrigeration or heating depending

A view of the stairwell-turned-cooler on Root Cellar Farm. Wooden brackets created flat spaces to place bins on the stairs, and the wooden platform (top) provided additional storage space.

on the time of year and weather. I used a CoolBot system during fall harvest when outdoor temperatures ranged from about 30 to 50°F (−1 to 10°C) at night to about 50 to 70°F (10 to 21°C) during the day. The concrete stairs, floor, and block walls in the cellar were uninsulated, and the first 5,000 BTUh air conditioner I tried couldn't reach the set temperature of 38°F (3°C). I then installed a larger 12,500 BTUh air conditioner with better results. Alongside the air conditioner, I installed a 6-inch (15.2 cm) duct fan (rated to 240 cubic feet per minute) to cool the space with outside air when possible (see "Cooling with Outdoor Air," page 116). The fan worked with two thermostat controllers, and it made the air conditioner unnecessary when the temperature was below about 34°F (1°C) outside, saving on electricity. The

The squash hammock at Root Cellar Farm in January. Onions and garlic were stored on pallets beneath the squash.

cellar needed cooling with outdoor air when temps ranged from 10 to 30°F (−12 to −1°C) outside, but below 10°F, the cellar needed supplemental heat. An inexpensive 1500-watt space heater plugged into a thermostat controller kept crops from freezing. I ran a fan at all times to circulate air through the cellar and keep temperatures even, and I always left air space between veggies and the walls to prevent freezing.

For much of the winter, the temperature in the stairwell hovered around 34°F, slightly too warm for root vegetables and cabbages but slightly too cold for potatoes. As a result, a portion of my carrots often began sprouting by late February and showing brown spots on their skin by mid-March.

In addition to the stairwell, I used a corner of the basement to store winter squash, onions, and garlic. I stretched some snow fencing over wooden framing to make flexible shelves, eventually dubbing this creation "the squash hammock." The squash hammock easily accommodated the roughly 600 to 800 pounds (272 to 363 kg) of squash I stored each year. Temperature control in this area was very low-tech. While the basement furnace incidentally heated the basement to between 55 and 60°F (13 and 16°C), I opened the two small sliding windows in that corner to keep the temperature there closer to 50°F (10°C). Onions and garlic sat on pallets on the floor under the squash hammock, where temperatures were in the mid-40s °F. This was technically too warm for the onions, which often started sprouting by mid-February, but the colder stairwell was too humid. The basement naturally stayed near 60 percent RH, just right for storing both squash and alliums.

Harvest and Processing

I harvested everything at Root Cellar Farm by hand and moved most crops around by wheelbarrow or sled (if harvesting after snowfall). I usually harvested directly into a set of plastic, vented, stackable totes that I purchased from a ginseng farm in Wisconsin. Because the yields were small, I could time the harvests as late as possible for each crop. For winter squash, this

Washing carrots on a spray table at Root Cellar Farm. Normally, I topped carrots in the field, but these carrots were for bunching and selling at a local fall farmers market.

meant harvesting before the first frost, whereas I harvested beets, celeriac, and cabbages before hard frosts arrived. For most everything else—except garlic and onions, which I harvested much earlier—I tried to minimize their time in storage. Weather at the farm was heavily influenced by nearby Lake Superior. As such, frosts were usually delayed until mid-October and hard freezes until mid-November. That said, October snowstorms weren't uncommon.

I tried to wash root crops before placing them in cold storage or, in the case of potatoes, curing. My washing system was simple and effective, albeit inefficient, with a simple hose sprayer and a spray table on sawhorses. I usually let crops dry overnight before placing them into their final storage bins or bags and moving them into the root cellar. Because I harvested late in the season when temperatures were cool, I could get away with storing veggies outdoors temporarily or in the unheated garage.

Of the crops I grew, four—garlic, onions, winter squash, and potatoes—needed some form of curing. I cured both garlic and onions in the garage on the property. This was a sheet metal building with a concrete floor, uninsulated walls, and access to electricity. At that time, I harvested and dried the entire plants for both garlic and onions. Onions dried in single layers on scraps of wooden lathe and fencing for four to six weeks under the gaze of a household pedestal fan. I tied garlic into bunches of four to five on both ends of short pieces of twine that I slung over some kind of support. One year, it was the exposed ceiling joists, and another I used an old swing-set frame. Garlic also took four to six weeks to properly dry. After cutting tops and loading them into mesh bags, I usually left onions and garlic in the garage until there was danger of freezing before moving them to their spot in the warmer basement. After harvest, potatoes went straight into the basement (where temps were in the mid-50s °F) in 50-pound (23 kg) burlap sacks for about a week under fans before going into cold storage.

Curing winter squash in the field was never really a possibility at Root Cellar Farm. By the time they reached full maturity and their stems were corking, there was often danger of frost and nighttime temps in the 30s °F. If the weather was relatively warm, I crammed all the squash and pumpkins onto a south-facing porch, where I exposed them to sun during the day and covered them with tarps or blankets at night. In years when the weather was too cold and fruits were in danger from chilling injury, I brought the squash directly to the squash hammock in the basement. There, I cordoned off the area with clear plastic—usually 6-mil polyethylene—and set a thermostat-controlled space heater and a fan to hold the squash near 80°F (27°C) with good airflow for about a week. After that time, I removed the plastic and opened the basement windows to bring the temperature down near 50°F (10°C).

Storage Containers and Locations

I used several types of storage containers depending on the crop. Potatoes were in plastic mesh or burlap bags stacked on the upper platform in the stairwell, where it was theoretically warmest. The potatoes still sweetened after they were stored around 34°F (1°C), but this usually went away if I brought them out into the warmer basement for two to three weeks. Cabbages were closed into waxed cardboard boxes in stacks of threes and fours. The winter CSA received kale and leeks early in the season—usually November—and the weather was usually conducive to leaving them covered in the field. One year there was an early cold snap with nighttime temperatures in the single digits, and I harvested kale and leeks into plastic bags and brought them into cold storage for a couple weeks without ill effects.

I stored most other root crops in vented stackable crates with plastic bin liners (to account for low humidity). These liner bags were unvented, and I controlled humidity in the bags by opening or closing their tops. This took close observation to avoid condensation and prevent veggies from drying out and becoming floppy. After attending a talk by Janaki Fisher-Merritt (see "Food Farm," page 219) at the

MOSES Organic Farming Conference, I tried lining bags with an absorbent material to absorb excess moisture while providing a source of humidity (emulating his method of placing a layer of peat moss covered by burlap at the bottom of each bin). The first year, I tried wood shavings but made the mistake of loading dry vegetables onto dry shavings. The dry wood shavings pulled moisture out of everything touching them, making the veggies on the bottom desiccated and floppy. Apart from the moisture problem, the shavings were messy and imparted a pine flavor to the veggies. In future years I used mixtures of brown paper bags and paper towels and loaded wet veggies directly into the bags with good results.

Marketing and Finances

I marketed about 80 percent of what I grew through a winter-only CSA that ran from early November through late March. Membership grew from eleven to nineteen families in the three years I ran the farm. Selling the CSA shares proved easy! Once word got out, the shares effectively sold themselves. The other 20 percent—excess beyond my needs for the CSA—sold primarily through a local grocery store. The produce manager was always excited to take whatever I could provide.

Being so small, the farm's gross annual sales were modest, growing from about $5,000 to $9,000 over the three years as I improved as a grower and adjusted my mix of crops. I made a profit each year despite startup expenses and equipment purchases. For most of the initial start-up costs, I used a zero-interest credit card that allowed me eighteen months to pay back my initial costs without fees. This essentially served as an operating loan, and I paid off all the initial expenses after selling CSA shares the first season.

Diseases and Pests

Three main pests affected crops on Root Cellar Farm: wireworms, cabbage loopers, and Colorado potato beetles. Wireworms are larvae of click beetles (family Elateridae), and they damaged root crops, especially carrots, in the first year. That's likely because the garden space was sod prior to plowing and tilling it into raised beds. After that first year, I saw few, if any, issues from wireworms. Cabbage looper larvae (*Trichoplusia ni*) are endemic to that region of the Upper Peninsula and happily chewed on any exposed Brassica crops. This was particularly problematic with cabbages, since customers were likely to find green worms hidden within the heads if I didn't mitigate the problem. I covered all my brassicas with floating row cover from transplanting/seeding until they were too large to fit under the covers. One year, this caused a fungal problem to fester on the cabbages, but it otherwise protected them from loopers. After that season, I removed row covers earlier without any issues from either fungi or loopers. The final pest, Colorado potato beetles (*Leptinotarsa decemlineata*), were a near-constant worry during the growing season. I controlled them through a combination of BT spray (an organic-approved solution of the bacteria *Bacillus thuringiensis*) and smushing the bright orange eggs by hand.

Advice for New Storage Farmers

There were two things about the storage and packing areas at Root Cellar Farm that I wouldn't recommend. Having the packing area in the basement meant I needed to carry veggies up a flight of stairs every time I made deliveries. That got old very quickly. The storage area itself was also a stairwell. The space wasn't cramped, but there was almost never a flat place to stand while lifting bins and boxes, and I had to navigate stairs while moving boxes in and out. Although I didn't suffer any serious injuries, it was obvious that lifting heavy objects on uneven footing was bad for my back, and I could have easily tripped and gotten hurt.

North Farm

Owner: Michigan State University

Location: Chatham, Michigan; about a half hour south of Marquette in the Upper Peninsula

Farm size: 5 acres in production with roughly 2½ acres dedicated to storage crops

Field setup: Bed system spaced to match tractor tire and skid steer width

Amount stored annually: 10,000 to 15,000 pounds (4,536 to 6,804 kg)

Signature storage crops: Carrots, cabbages, onions, beets, garlic, and winter squash

Winter markets: Mostly wholesale, with about 98 percent going through one local food co-op

Summer markets: Same as winter

Portion of annual sales from storage: Roughly 40 to 50 percent

Number of people working: One farm manager plus several part-time workers, totaling roughly two to three full-time people through the summer and fall

Michigan State University's North Farm is roughly a two-hour drive from Root Cellar Farm. While living in the Upper Peninsula, I became friends with the previous production manager, Allison Stawara. I didn't know anyone else storing winter crops in the area, so Allison and the North Farm became great resources for sanity-checking my methods and trading notes. Long-term winter storage has been at the core of North Farm's production since returning to vegetables in 2014, and a lot of experts in the field of northern vegetable production have contributed to the farm's design and systems. That's why I was excited to chat with Sarah Hayward, the current production manager, to hear details about their methods and the current state of winter storage at MSU's North Farm.

Journey to Wintertime Storage

The North Farm is part of Michigan State University's Upper Peninsula Research and Extension Center. The farm was established in 1899 as a place for Michigan

A harvest-time sunrise at the MSU North Farm. *Photo courtesy of Sarah Hayward.*

State researchers to study dairy production, orchard crops, short-season vegetables, and even draft-horse breeding. In 2014, North Farm received a USDA Agriculture and Food Research Initiative (AFRI) grant to revitalize its vegetable research and production. The farm staff now conduct variety trials, provide educational outreach, and host an incubator program for beginning farmers. Storage crops fit nicely into the farm's programming because they take relatively little maintenance during the growing season and allow staff to focus on research for much of the summer.

Storage Spaces

The farm's legacy of dairy farming eased the transition to vegetable storage, providing many outbuildings for storing and processing crops with only minimal

adjustments. The primary storage area at the North Farm is a root cellar in the basement of an old barn used for dairy research. The space is roughly 24 by 30 feet (7.3 by 9.1 m), but two rows of central concrete pillars limit both storage space and mobility. The walls, arching ceiling, and floor are made of poured concrete, and the cellar is earthed in on three sides. On the access wall, North Farm staff added 2 inches (5.1 cm) of EPS foam board onto exterior concrete for added insulation.

Although they experimented with mechanically cooling the space, the root cellar now maintains storage temperatures mostly by passive means. North Farm staff originally tried using a CoolBot system, but they found it difficult and energy-intensive to keep the cellar cold enough. Since neither the slab nor the

The retired refrigerated trailer used as temporary cold storage during fall harvest at the North Farm. *Photo courtesy of Allison Stawara.*

178

earthed-in walls are insulated, the cooling equipment fought with the large thermal masses of concrete and the soil beyond, and it was a losing battle. Rather than making expensive changes to the cellar, the North Farm purchased a retired refrigerated trailer to serve as temporary cold storage from harvest time until the root cellar naturally reaches cold-storage temps.

The trailer's original refrigeration system no longer functions, so they cool the space with two air conditioners and CoolBots. They're able to monitor temperatures in the trailer remotely with wireless temperature probes, which send text alerts to the farm manager's cell phone if temperatures go below 32°F or above 40°F (4°C). As the weather turns colder, and especially at night, North Farm staff use fans to blow cool air into the root cellar to accelerate

The root cellar at the North Farm filled with wooden pallet bins. This is also where crews pack veggies for orders. *Photo courtesy of Sarah Hayward.*

its cooldown. They have a general goal of keeping crops in the trailer for no more than a month, and they're usually able to move veggies out of the trailer by mid-to-late November, when temperatures in the cellar drop below 40°F. Eventually, the root cellar decreases to about 35°F (2°C), where it stays consistently until around the end of March.

The North Farm has a second storage space in the basement below their staff offices, in a building they call the Holstein Hilton. There, conditions are both warmer (about 50°F, or 10°C) and drier than the root cellar, and they use the space as long-term storage for winter squash and garlic.

The North Farm does not control for humidity in either of its storage spaces and instead relies upon bin liners to create humid microenvironments for the crops that need high humidity. This technique also allows them to store onions—which need cold and dry conditions—in the root cellar alongside roots and cabbages. Although humidity is not monitored in the root cellar, it tends to be a damp place with moisture occasionally seeping through the walls. To maintain good air circulation and prevent cold spots inside the cellar, they run box fans hung from the ceiling for most of the winter.

Materials Handling and Packing

Storage containers in North Farm's root cellar are stackable wooden pallet bins that are compatible with standard pallet jacks and large enough to hold about 600 pounds (272 kg) of carrots when full. The crew pressure washes the bins each fall and also fits them with new polyethylene bin liners depending on the crop. In the Holstein Hilton, they use single-use cardboard Gaylord bins (see "Storage Containers," page 47) for winter squash and poly-mesh bags for garlic.

The North Farm uses a skid steer outfitted with forks when moving veggies between various storage and washing areas, but there are challenges, especially during the winter. One problem is that the entrance to the root cellar is too small for the skid-steer, and the uneven cellar floor makes moving bins with pallet jacks a pain. Another complication is snow. Chatham receives 150 to 200 inches (3.8 to 5.1 m) of snow per year. Both the refrigerated trailer and the washing shed are more than 100 feet (30.5 m) from the root cellar, and clearing paths can be a laborious task. Sometimes North Farm staff decide to move veggies from the trailer to the cellar early to avoid the mess of early, heavy snowfalls. Whenever this happens, they run fans at the door of the cellar to blow cold, outdoor air inside to keep veggies within a food-safe temperature range. The distance between cellar and wash shed is one reason why the North Farm packs veggies for sale inside the root cellar. (Another is the need to heat the washing shed with space heaters while working there in the winter.)

Harvesting and Processing

North Farm staff try to hold crops in the field as long as possible to minimize their time in storage and to run the refrigerated trailer as little as possible. That said, they're sometimes limited by the availability of fall help, which often comes from college students. Sarah said they're sometimes forced to harvest whenever workers are available rather than when it's optimal for crops.

Garlic and onions are always the first storage crops to come out. The crew hand-harvests garlic onto a flat hay wagon, then hangs the plants to dry under large fans until the hardneck stems are no longer moist when cut. After that, they grade the garlic with a homemade wooden tool consisting of two boards fastened in a wedge pattern onto a larger plank. They sort garlic into size grades based on how far bulbs fit into the wedge. After sorting to commercial grades, they store garlic in poly-mesh bags in the basement of the Holstein Hilton.

Ideally, the crew waits to harvest onions until a majority of tops have fallen, but, for labor reasons, they can't always wait. Sarah shared her trick of using a powerful leaf blower to crimp onion tops and force early maturation, something she did last season to align the harvest with the availability of help. After crimping, she let the onions dry in the field for three to four days before harvesting and moving them to the greenhouse to cure. Sarah admitted this trick was a gamble but said it worked out swimmingly. Onions cure on wire racks in the propagation greenhouse (with a concrete floor) for two to three weeks with the end-wall exhaust fans running at full blast. They elevate the racks on cinder blocks, allowing onion tops to dangle beneath while bulbs and roots get more sun exposure and airflow. Sarah also mentioned that shade cloth over the greenhouse is important to prevent the onions from sun scalding. She said that hand-cleaning onions can be a lengthy process and is something they do whenever there's time in the fall. The goal is to fully clean and trim onions, get them into wooden bulk bins, and move them to the root cellar before freezing temperatures take over the greenhouse.

The crew harvests winter squash after onions but before frosts arrive, usually in early to mid-September. Sarah said there's usually time for some field curing while squash are still on the vines. Once out of the field, winter squash take over a portion of the same greenhouse occupied by the onions to cure. The crew lays out squash in a single layer and pulls back enough shade cloth to expose squash to the sun. After several weeks and before nighttime temperatures drop to near freezing, they separate squash by variety, load them into cardboard Gaylord bins, and move them to the Holstein Hilton for long-term storage.

Harvest timing for the final three storage crops—carrots, beets, and cabbage—depends on the weather and pest pressure from rodents. They try to avoid exposing beets and cabbages to hard frosts and chiseling carrots from frozen ground. If the rodent pressure is low, Sarah prefers that carrots stay in the ground as late as possible, and she said it's common to harvest the last carrots from under a light layer of snow.

The crew at the North Farm get some soil-loosening and materials-handling help from the tractor and skid steer. They typically run the undercutter bar with their tractor to lift beds with beets and carrots prior to harvest. They're also able to drive pallet bins into the field with the skid steer. This way, they can harvest directly into large bins rather than transferring from smaller containers later. With the skid steer, they load full bins into the refrigerated trailer and use pallet jacks to move bins to their final positions once inside.

Most trimming occurs in the field alongside the harvest. They usually break carrot tops by hand but trim beet tops and cabbages with harvest knives. Sarah mentioned that they intentionally leave the cabbage stems extra-long to minimize losses when trimming off outer leaf layers after storage. To minimize disease and pest problems, they try to remove all cabbage residues from the field before the snow falls whenever the workload and weather allow. Carrot and beet residues stay in the field, where they're tilled under the following spring. The crew field-culls crops with aesthetic problems and removes pieces affected by disease or insects. Many field culls go into separate bins and are given to local farm workers and others to use as deer bait.

The crew prepares the wooden storage bins differently depending on the crop. Unwashed root crops like carrots and beets get fresh burlap on the bottom of each bin and a nonperforated bin liner placed upside down over the bins. Sarah also mentioned that they pour water into bins of dirty carrots to moisten the burlap and increase the overall humidity. Without this step, they've noticed worsened soil staining and skins with rougher textures after months in storage. With cabbages, they put fresh, nonperforated liners inside the bins prior to harvest, and they loosely fold over liner flaps when placing bins in the cooler. They don't use liners with onions, but they place a closed wax box in the bottom center of each bin. According to Sarah, the box helps to reduce the elevated temperatures that occur as the onions respire and produce ethylene gas.

North Farm washes root crops in a custom-made barrel washer as the crew prepares wholesale orders. (Although, if there are only a few hundred pounds of something—for example, last season's beets—they might wash these before long-term storage.) Because

Carrots in wooden bins with bin liners placed over the tops to hold in moisture. *Photo courtesy of Sarah Hayward.*

they have the facilities to run their root barrel washer during the winter, they feel less inclined to wash root crops in the fall. This way, they're able to spread the workload of washing over the winter rather than cramming it into the fall when they're short-staffed.

Sarah said it's most efficient to have two people running the barrel washer: one person loading veggies and one spraying roots with a pressure washer. The crew hand-loads dirty roots from bulk bins into grain shovels to transfer into the washer, which can handle between 100 and 200 pounds (45 to 91 kg) at a time. After tumbling for several minutes and passing inspection, cleaned roots drop directly into bulb crates to drip dry. The clean roots usually stay in the heated washing shed overnight to dry before they're transferred to a lined wooden bin with a clean liner (with flaps loosely folded over) and moved back to the root cellar. Carrots are the main root crop that require washing through the winter, and staff typically wash one bin every other week to maintain a constant supply of clean carrots for packing orders.

Marketing

Despite the farm's affiliation with MSU and its focus on research, commercial vegetable production is still an important part of the North Farm's programming and overall budget. That said, the North Farm does its best to avoid competing with or undercutting other local farmers, and winter storage and sales is a key part of this strategy. The North Farm's goal is to fill their community's need for local vegetables as long as possible into the winter after other farms have run out of produce. They emphasize premium quality in their products to get a high price point, maintaining both USDA Organic and Good Agricultural Practices (GAP) food safety certifications.

Sarah estimates they sell 98 percent of their crops through the local food co-op in nearby Marquette, with the most sales happening in January. The gross annual sales from storage crops are around $35,000. Carrots are by far their most popular crop, and they sell them in bulk (loose), in 2-pound (1 kg) bags, and in 5-pound (2.3 kg) bags. They make some bags with assorted color varieties, but most of their sales are from orange carrots. They sell all their other storage crops in bulk, which requires less time packing in the root cellar.

The packing area in the root cellar consists of a table with a precision scale. Depending on the crop and the order, the crew either packs into waxed boxes with single-use liners or weighs into bags. For bagged crops, the North Farm prefers nonperforated bags with twist ties.

Sarah explained that spoilage rates in storage sometimes depend on pricing strategies. When they set higher prices, the crops sell more slowly, and losses become more likely. If, for example, North Farm asks for higher prices with cabbages, they have to store cabbages through March and can lose up to 30 percent to spoilage. Selling at lower prices usually leads to faster sales and fewer losses. We agreed that it's difficult to know which strategy is best and what price will maximize income while minimizing storage losses.

Favorite Varieties

North Farm has a few stand-out favorites from recent trials and production. Bolero carrots have performed well, with consistency in both size and shape. Storage No. 4 and Passat are favorite green cabbage varieties. Favorite storage onions include Red Carpet and Talon (yellow).

Advice for New Storage Farmers

Sarah's advice is to make a schedule for planting, maintaining, and especially harvesting and processing crops for storage. "Make a schedule, make a plan, and stick to it," she said, emphasizing how stressful it is to fall behind. Adding my own two cents to Sarah's advice, you should be conservative when planning. Developing a sense for how long tasks require takes time and experience. Keep good notes and records, make reasonable estimates, and then add in extra time to be safe.

Offbeet Farm

Owners: Sam and Danielle Knapp

Location: Fairbanks, Alaska; about 10 miles west of town

Farm size: 1 acre devoted entirely to storage crops

Field setup: Permanent raised beds with 30-inch (76 cm) beds and 18-inch (46 cm) walking paths

Amount stored annually: 25,000 pounds (11,340 kg)

Signature storage crops: Carrots are by far the most popular, but the CSA members also receive beets, garlic, onions, cabbage, kale, kohlrabi, rutabagas, radishes, turnips, parsnips, and winter squash.

Winter markets: Winter-only CSA (one hundred families) and occasional wholesale deliveries

Summer markets: With the exception of garlic scapes, no summertime sales

Portion of annual sales from storage: Over 99 percent

Number of people working: Sam full-time in summer and part-time in winter, and two full-time helpers during harvest (four to five weeks)

Offbeet Farm is my current (hopefully long-term) farm near Fairbanks, Alaska. From the beginning, I've focused the farm on storage crops and wintertime sales, and the farm demonstrates that it's possible to start a successful storage-focused farm from scratch in only a few years. Despite relatively steep upfront costs, Offbeet Farm has thrived from the beginning and managed to quickly build a following of customers with its winter-only offerings. Farming in Alaska is difficult for many reasons—short growing seasons, expensive supplies with limited availability, moose—but it's also a ripe place for localized farm businesses. Alaskans, especially in the interior, understand scarcity and are acutely aware of their precarious position at the end of the supply chain. Because the vast majority of our food gets shipped in, the community highly values its local producers.

Journey to Wintertime Storage

I started building Offbeet Farm in 2020 after moving with my wife, Danielle, from the Upper Peninsula of Michigan. It wasn't a smooth process; there were several stops and starts as we dealt with hurdles related to financing and land clearing. We bought a house on three sloped and wooded acres outside Fairbanks, largely because the property had a well, which is rare in the area. We cleared the trees from roughly half the property toward the end of the first summer.

As I'd hoped, storage farming worked just as well in Alaska as in Michigan. I knew how to grow, store, and market storage crops from my time at Root Cellar Farm, and I had a basic understanding of the farming community around Fairbanks from a summer spent there as a farmhand. I also knew that because of limited income during startup and high upfront costs, I'd have difficulty hiring help right away. Storage farming spread some of the typical summertime workload to the offseason months, allowing me to grow more over a larger area than I ever could have done alone on a typical veggie farm. As in Michigan, word quickly spread about the farm. It wasn't long before the demand for my produce quickly outstripped the supply.

Storage Spaces

Financing and building the storage building at Offbeet Farm was a saga in itself. My local Farm Service Agency (FSA) office was severely short-staffed during the height of the pandemic, and after an unusually long wait, I learned that my first farm-loan application had been rejected. Rebounding, I redrew the building designs, parted with my contractor, and managed to form a budget that squeaked in under the $50,000 limit for the FSA's microloan program (which requires less money upfront and less collateral to backup loans). After a long period of consideration,

Sam and the crew harvest beets into 5-gallon buckets on Offbeet Farm. *Photo courtesy of Phil Knapp.*

the FSA approved this second application in late June of 2021. That left me with roughly three months to go from a pile of dirt and stumps to a functional storage building to house that year's harvest. (The farm wasn't yet at full capacity, with about ¼ acre in production that year.) The following three months were some of the most grueling yet exciting and rewarding of my life. I couldn't have finished the project without help from my friends, family, and the manic energy that endless daylight bestows on residents of the far north. The storage building at Offbeet Farm is built into the hillside with an overhead door and an entry door at the front of the building. I wish the building were larger, but given the limitations we had financing the construction, the building meets our needs remarkably well. The footprint is a square measuring just under 24 feet (7.3 m) wide on the interior. An insulated (2 × 6 framing) wall divides the space down the middle into a cold-storage room and a warm-storage/packing room. A 5-foot-wide (1.5 m) insulated sliding door separates the two spaces.

During harvest season, I keep the cold-storage room around 36°F (2°C) with two CoolBot systems working in tandem. One window air conditioner has 24,500 BTUh of cooling power, while the other has 12,000 BTUh; this setup is a bit underpowered, and I plan to add an additional large air conditioner for next season. I also keep the humidity in that room as high as possible during the harvest, usually around 90 to 95 percent relative humidity. If I go higher, the air conditioners tend to ice up. There's a balancing act between keeping the space both cold enough and humid enough. If I notice any desiccation in the veggies, I'll use scrap pieces of polyethylene sheeting to temporarily cover any open containers and retain moisture.

Once outdoor temperatures drop consistently below freezing, I turn off the Coolbot systems and switch to a cold-air intake fan. This is an inline duct fan plugged into two thermostat controllers connected in series (see "Cooling with Outdoor Air," page 116). When the cold-storage room is full of veggies, the cooling system needs to be active when outdoor temperatures are between 32°F (0°C) and about −10°F (−23°C). The building has enough insulation and the veggies create enough heat through respiration that the cooling fan has to run periodically despite the frigid outdoor temperatures. Between −10 and −20°F (−23 and −29°C), the storage room (when full) hovers around freezing without supplemental heat. The room sometimes needs supplemental heat when it starts emptying out later into winter, but a small 1,500-watt space heater plugged into a thermostat controller is enough to keep it above freezing. Even in a near-empty storage room, the heater only runs when the outdoor temps drop below 0°F (−18°C).

Humidity in the cold-storage room hovers between 96 and 99 percent for most of the winter after the CoolBot systems are off. I use a commercial centrifugal humidifier (Active Air brand, see resources on page 237) that's gravity fed from a repurposed 5-gallon brewing bucket. A digital humidistat controller monitors humidity and turns the unit on and off as needed. When it's relatively warm outside, the intake fan runs more and brings in more dry air compared to when it's colder. At the beginning of the storage season, I need to refill the 5-gallon bucket every three to four days. Later in the season, one to two weeks might pass before I need to add water.

I store crops that need lower humidity—winter squash and pumpkins, garlic, and onions—on the "warm" packing side of the building. As long as winter squash are present, I keep the room at 50°F (10°C). Humidity mostly takes care of itself, staying between 50 and 60 percent. It's a little on the dry side, but I think it's helpful for storing *C. maxima* squash, which tend to have thick peduncles that don't always dry out during the short growing season. Neither storage onions nor garlic grow particularly well around Fairbanks, so I don't grow large quantities of either. I usually run out by mid-January, before the warm storage temps cause early sprouting. After I run out of winter squash (usually in January), I lower the packing room temperature to between 40 and 45°F (4 and 7°C) to save on heating costs.

My original heating system was a ceiling-mounted propane heater, but I've switched to an electric resistance heater for a few reasons. First, the propane heater couldn't hold a tight temperature range, and the room fluctuated by 10°F (6°C) or more during heating cycles. More importantly, I didn't trust the heater to work 100 percent of the time. This model had a pressure switch designed to prevent back-drafting of exhaust, but the rubber tubing feeding this sensor was easily clogged by soot and condensation. When the tubing clogged, the heater didn't work. Although electricity is expensive, the new electric heater is more reliable, more efficient, and can hold a much tighter temperature range with its digital thermostat.

Materials Handling and Containers

Finding affordable and functional storage containers and ways of moving veggies around the farm have been some of my biggest challenges, exacerbated by the farm's location in Alaska. It's difficult to find containers and lifting equipment locally, and shipping them here is often prohibitively expensive. That said, I think new storage farms anywhere will face these same challenges to varying degrees.

During harvest, processing, and storage, all the crops on the farm are lifted by hand. As such, I've thought of ways to limit bodily strain for the crew and me. For all crops except winter squash and cabbages, 5-gallon buckets are my preferred harvest containers. Their small size limits how much the crew and I have to lift—one bucket holds 20 to 25 pounds (9.1 to 11.3 kg) of most crops—they're easy to lift ergonomically, and you can balance the load by carrying two at a time. For the bulkier cabbages and winter squash, I use either 27-gallon totes or stackable crates that are roughly the same size. These containers can be heavy when full, so I encourage the crew (and myself) to either partially fill them or have a buddy help lift them.

Moving crops from the field to the washing station happens in a few ways. Most often, we use a rickshaw-style cart that holds up to ten buckets at a time. It can be dicey rolling a fully loaded cart on a downhill path, so we often leave the cart on the relatively flat central path and schlep buckets to it. However, the cart can fully straddle the raised beds with its bicycle tires if necessary. We can also use a cart pulled by the BCS two-wheeled tractor, but we rarely do this. The hand cart is faster and more maneuverable, and it can hold more buckets than the current BCS cart. (I'm considering upgrading to a larger flatbed cart to make the BCS setup more efficient.) A third option is to drive a pickup truck onto the field's central path. This works great for hauling larger quantities at once, such as cabbages and squash. This isn't an option if it's been raining, because the truck can easily slip on the path. For better or worse, I valued our limited growing space over large-vehicle access in the highly sloped field, and getting the pickup truck in and out feels uncomfortably tight at times.

From the washing station (more on that later), crops destined for cold storage are dropped into vented bins 40 to 50 pounds (18.1–22.7 kg) at a time. We roll several bins into the cold room on a wheeled cart and weigh them before unloading the veggies into their final containers. For crops and varieties that we store a lot of—orange carrots and red beets, for example—I've chosen woven polypropylene super sacks as storage containers because they're inexpensive and relatively space efficient; one super sack can easily hold 1,000 pounds (453.6 kg) of carrots. The super sacks sit on pallets, and stacks of two are possible using pallet racking made from wooden beams. I line the bottom of the sacks with a thin layer of peat moss covered by clean pieces of old floating row cover. Because the farm doesn't have any heavy handling equipment yet, we have to load (and unload) the sacks in place, which is admittedly a pain. Although the super sacks are

entirely open on top, I fear that crops deep inside the bags don't always get enough airflow. I'm thinking about cutting small ventilation holes in the bags to help, but I've already stopped storing kohlrabi in the super sacks since they need more airflow. Eventually, I want to upgrade to custom wooden bins and a hydraulic pallet stacker to make loading and unloading more efficient and to make better use of the space.

I store specialty crops in smaller amounts in plastic 27-gallon totes. I line the bottom of each tote with covered peat moss, and I've peppered the sides, tops, and bottoms with ventilation holes. These totes weigh around 100 pounds (45.4 kg) when full and comfortably stack about five high. Despite the ventilation holes, condensation still builds up inside the containers, and I prefer to leave the lids off whenever possible.

When packing orders, I rarely pack more than 300 to 400 pounds (136–181 kg) of any one crop at a time. As such, I don't mind unloading veggies into empty totes by hand and wheeling them around on a strong, table-height cart. I can wheel a couple hundred pounds from the storage room to the packing room without ever lifting the heavy loads. To access the upper stack of super sacks, I use a compact rolling scaffold, which affords me a spacious and sturdy platform from which to work. Again, the system isn't perfect and feels inefficient at times, but it has allowed me to store and process roughly 25,000 pounds (11,340 kg) annually without special lifting equipment or expensive containers.

Harvest and Processing

There's an ominous feeling to harvest season in Alaska that I haven't felt anywhere else. I think it's because the transition from summer into winter is so abrupt. Mean daily temperatures drop from 52°F (11°C) on September 1st to 38°F (3°C) by October 1st; by November 1st, the daily mean is just 13°F (−16°C). I'm acutely aware that the ground will likely be frozen by mid-October with more-frigid temps to follow.

Both garlic and onions usually come out before the true rush of harvest season, throughout the month of

August. I don't grow much garlic and can usually harvest and process my crop in an afternoon. Most bulbs require some forking to pull out without breaking. I've recently started washing soil off my garlic roots before drying them, and I find that garlic dries easily. I like to clip the stems to about 6 inches (15.2 cm) before stowing them into dense single layers in vented and stackable bins for curing. These go under fans in the attic of the storage building, where they remain until temperatures threaten to freeze the garlic, usually in late September or early October. Whatever garlic I don't replant, I put into burlap sacks and bring into the packing side of the storage building.

Storage onions are a tricky crop in interior Alaska. August can be a wet and chilly month in Fairbanks, so I try not to leave mature onions in the field too long for fear of fungal rots entering through the necks. When harvesting, I rip off tops to 4 to 6 inches (10.2 to 15.2 cm) in the field in an attempt to leave some organic matter behind. Harvested onions go immediately into the attic of the storage building, where I spread them in single layers on wire racks to dry under fans for three to six weeks. Once they are dry, I fill vented crates in the packing room with the dirty onions and clean them as I need them. I have aspirations of bringing nothing but cleaned onions inside, but, alas, there always seem to be more pressing tasks in the fall.

The official harvest season, when I hire a crew of two to three people, starts at the beginning of September. During this time, we're focused entirely on harvesting, washing, and getting crops into storage until it's all done. Harvest order can vary, but I'm generally harvesting things like winter squash, beets, and turnips first, while leaving crops like carrots and parsnips for last.

Crew member at Offbeet Farm cleans beets in the custom barrel washer. *Photo courtesy of Phil Knapp.*

I harvest winter squash and pumpkins within a few days of mean temps dropping below 50°F (10°C) or before the first frost, whichever happens first. Ideally, there will be a few days of sunny, dry weather during which the squash can begin curing and drying their stems. I try to harvest them only when it's dry and to avoid rain any time after the fruits are cut. When I deem it's time to bring them inside—due to temperatures or approaching rain—we set the squash in single layers on well-vented racks and in vented bins. I cordon off the back half of the packing room with a polyethylene sheet and set the thermostat to 80°F (27°C). The squash and pumpkins cure for at least a week with a dehumidifier set to 55 to 60 percent relative humidity to avoid mold formation and help suck moisture out of the thick *C. maxima* stems. After curing, I lower the thermostat to 50°F and keep the dehumidifier running. There are usually a few individual fruits that go bad either during or shortly after curing, and it's important to give the crop a couple of close inspections to remove any bad actors.

We harvest all of the remaining crops in similar ways, sending them through the barrel washer prior to storage. The exceptions are cabbages, which go unwashed, and kohlrabi, which we spray off in vented crates. All trimming occurs in the field as a means of increasing the soil organic matter on the raised beds. I'm a fan of trimming everything with small, sharp harvest knives, except for some varieties of carrots with thinner tops, which we break by hand. Ideally, we wash crops as we harvest, and I've found that one person at our wash station can usually keep up with two to three people harvesting, depending on the crop.

After washing, we use a semicircular paddle on a long handle to pull the veggies from the barrel washer onto a grading table. Our initial grading simply removes unsalable or broken pieces or those likely to rot in storage. These get discarded into an unmanaged pile or go home with crew members or friends. From there, veggies get pushed down a ramp into a waiting bin sitting on a wheeled cart, which we use to roll veggies into the cold room.

Kale is usually the last vegetable out of the garden, since I try to leave it unharvested until the last possible moment. I don't mind if kale goes through several hard frosts, but I don't want the leaves to freeze solid. That said, I minimize its time in storage because losses to yellowing can be so high. The youngest kale leaves, toward the top of the plants, can stay green and flavorful for remarkably long times in storage—I've seen the uppermost leaves go more than four months—but the more mature leaves won't last nearly as long. After a few years of observations, my current strategy is to market the mature kale leaves—those showing signs of flattening and unfurling, which is easiest to spot on curly kales—immediately, while keeping the less-mature leaves up to eight weeks for either the CSA or farmers markets. These leaves are packed loosely into a super sack. If I pack them too tightly, the leaves both suffocate and build up ethylene, resulting in yellowing and bitter flavor.

Marketing

Apart from garlic scapes, I don't sell anything during the primary growing season. This is an intentional choice; I certainly could grow things for the hoppin' Tanana Valley Farmers Market in Fairbanks, but I choose to devote my time and energy to tending my storage crops. The first sales of the storage season usually wait until after the harvest is finished in early October. These might be extra kale or a squash variety that isn't holding well in storage, and I usually reach out to my wholesale partners or my CSA list.

About 80 percent of the farm's gross income comes through the winter CSA. Between the two share sizes, I have about one hundred families enrolled. The first shares start at the beginning of November, and members receive veggies every other week through mid-to-late March. Potatoes for the CSA come from another Fairbanks farmer, who sells me bulk potatoes at wholesale price.

Unfortunately, the Tanana Valley Farmers Market concludes in late September, so I don't reach much of the hungry farmers-market crowd. I participate in two markets during the holidays: one relatively small pop-up market before Thanksgiving, and a holiday bazaar put on by the farmers market association. The holiday bazaar is a huge day for the farm both in terms of income and marketing. Since it's in mid-December, I'm usually the only of 200 or so vendors with a full array of produce, which makes my booth a popular stop for marketgoers.

Sales to my two wholesale partners make up about 10 to 15 percent of annual income, though I heavily favor one of these vendors over the other. Right around the time I was building Offbeet Farm, another local business called the Roaming Root Cellar was also getting off the ground. Their original idea was to have a bus acting as a mobile retail space to help local food and art producers sell their wares. Unfortunately, the bus bit the dust fairly quickly, but the store rented a retail space, and we have grown our businesses together. I'm the main supplier of winter veggies for the store from interior Alaska; whatever they can't get from me, they get from an Anchorage-based food hub. Truthfully, I wish I could supply the store with more, but the reality of running a farm on small acreage is that I need to direct-market most of my veggies to make a living. Still, it's an option for potential future expansion.

Throughout the winter, I grade produce as I'm packing orders for the CSA, markets, and wholesale. "Seconds" are veggies that are broken or oddly shaped, have weird blemishes, or require heavy trimming. I set these pieces aside and either donate them to a local soup kitchen or market them at half price. I also set aside small pieces, especially carrots and beets, and sell these in "mini" bags through my wholesale partners. Customers love tiny things if you label them as such! I collect any pieces that are unsalable or spoiled and give those to a local pig farmer who's willing to pick up from the farm.

Diseases and Pests

So far, I haven't seen many of the common vegetable diseases besides environmental molds on Offbeet Farm. Pests, both vertebrates and invertebrates, are the main challenges. Moose pose an enormous threat to my livelihood; a single moose can cause thousands of dollars of damage to fences and crops in just a few hours. They gladly take large bites out of cabbages and trample root crops while browsing. I use an 8-foot electrified fence around the entire farm to keep them out. While a simple fence won't stop a determined moose, the key is deterrence and avoiding habituation. If a moose gets in once, good luck stopping it from busting its way in again.

Cabbage- and onion-root maggots are other problems. I use row cover to protect my crops from the flies during the growing season and fall tillage to disrupt many of the pupae from overwintering.

Ironically, meadow voles cause much more damage each year than moose. Storage crops are among their favorites, and if left unchecked, they can ruin entire plantings. I've seen them chew out gaping holes from squash, burrow through beets and carrots, and—in the most aggravating way possible—take a single bite from each beet in a row. To cut their numbers, I do two things. First, I have cats that are constantly going after voles in and around the field. Second, I deploy a small army of traps from the beginning of the growing season until after harvest. I use plastic Victor rat traps that you can set with one hand, and these go under open-bottom boxes I made from treated plywood. The boxes have semicircular holes cut from the edge contacting the ground, and voles scurry under the openings and into the traps. I adopted the boxes to avoid killing birds and to protect my cats; so far so good.

Favorite Varieties

Upon moving to Alaska, I wasn't sure if the varieties I grew in Michigan would perform similarly. After a few years of trying, I've landed upon some favorites that seem to perform well no matter the variations in summer weather patterns. For carrots, two stand-out varieties are White Satin (white carrots) and Purple Elite (purple skin with yellow cores). Both grow extremely uniformly, taste amazing, and reliably last for four months in storage. My favorite winter squash is Sweet Mama, which is a dark green kabocha-type with a compact habit. Small Sugar pie pumpkins have remarkable yields and reliably turn bright orange by late August. Lil' Pump-Ke-Mon is a customer-favorite pumpkin that tastes great and is also beautiful, like an ornamental. Kalibos is a beautiful red pointed cabbage that stores well and that customers love. I find it sizes up more reliably than the round red cabbages.

Advice for New Storage Farmers

I advise farmers new to winter storage to be pessimistic in their planning—a practical version of plan for the worst, hope for the best. This goes for both crop yields and losses in storage. I use my past yields to project future harvests, but I tend to keep my estimations on the low end. Similarly, I overestimate how much of each vegetable I'll lose in storage. Put another

way, I tend to underestimate the salable weight I'll get from each crop. These two practices help ensure that I have enough inventory to meet my obligations to the CSA and the expectations of my wholesale partners. Losses are an inevitable part of storage farming, and it's comforting to think, "It's ok; I planned for this," when you're faced with spoiled veggies.

I also advise new storage farmers to weigh or measure veggies at the beginning of the storage season—in other words, know your starting inventory. With this initial inventory, you can plan sales amounts to each customer while holding back some to account for losses. I've never found it practical to precisely track spoilage weights—I don't want to weigh a rotten, dripping pumpkin—but I do track the salable weights that go out the door. This way, it's possible to know when to stop selling cabbages to my wholesale partner, for example, to ensure that I can fulfill obligations to the CSA.

The Offbeet Farm harvest crew loads Lil' Pump-Ke-Mon pumpkins onto the harvest cart. *Photo courtesy of Phil Knapp.*

Jonathan's Farm

Owner: Jonathan Stevens

Location: Winnipeg, Manitoba, Canada; about 10 miles north of town

Farm size: 6 acres in production with just under 2 acres dedicated to storage crops

Field setup: Bed system spaced for tractor tires (roughly 5 feet, or 1.5 m)

Amount stored annually: 40,000 to 50,000 pounds (18,144–22,680 kg)

Signature storage crops: The three storage staples are carrots, onions, and potatoes, but the farm also sells squash, cabbages, beets, celeriac, kohlrabi, turnips, and radishes in the winter

Winter markets: A free-choice winter CSA with 280 members, occasional wholesale

Summer markets: A 500+ member CSA, weekly farmers market, and occasional wholesale

Portion of annual sales from storage: 16 percent

Number of people working: Jonathan full-time plus seven full-time crew members in the summer and fall; a few employees offer part-time packing help in the winter

Jonathan's Farm is an example of a thriving, summer-oriented market farm that has added a small but meaningful storage component to their business. Winter storage has given the farm an additional income stream without adding too much to the summer workload, and it's helped them retain good employees through the winter season. For owner Jonathan Stevens, winter storage was an obvious opportunity to improve and diversify the farm. He's managed to expand into winter storage with relatively little specific investment, utilizing many of the same tools and infrastructure he uses for summer production. Jonathan's Farm is the only farm I've encountered with a free-choice winter CSA. During our interview, he talked me through the logistics of the CSA and how he plans for each season. In light of our discussion, I'm considering switching over to the free-choice CSA model myself.

Journey to Wintertime Storage

Jonathan started Jonathan's Farm on rented land in 2010. There, he focused mainly on establishing a summer-oriented CSA and market farm, building both experience and capital. Jonathan first dabbled with the winter CSA in 2014, in his last season on rented land. That year, he built a small, insulated room in the basement of his house in Winnipeg to keep a small amount of storage crops. He kept the storage room cold by cracking a window and monitoring the temperature with a remote thermometer. The storage containers were small and simple, either woven poly sacks or mesh bags. Buoyed by success

The crew at Jonathan's Farm harvests cabbages on a sunny fall day. *Photo courtesy of Jonathan Stevens.*

and customer demand, Jonathan continued offering the winter CSA the following year after moving his operation to newly purchased farmland. He continued to store winter crops in his basement until 2017, when he built a larger wash-pack shed and cooler at the new farm. Jonathan's wintertime sales have helped him retain some summer customers, access new customers who keep their own summer gardens, keep some employees on during the winter, and improve the overall finances on the farm.

Storage Spaces

Jonathan's Farm currently uses two walk-in coolers as winter storage spaces. Both have roughly 250 square feet (23 m²) of floor space with 7-foot (2 m) ceilings. Jonathan built the first cooler inside the building used for washing and packing by adding two interior walls and a shorter ceiling to a back corner—I'll refer to this one as the primary cooler. These walls and the ceiling were built with 2 × 6 framing and fiberglass insulation sandwiched between foam board covering both sides; the total R-value is roughly R35. Jonathan built the second cooler as an addition to the wash-pack building. This second cooler gives him the flexibility to harvest certain storage crops earlier while the primary cooler is still full of summer veggies. The new concrete slab was insulated on the sides but not beneath, and the walls and ceiling were framed and insulated in a similar manner to the first cooler. Both coolers are used in the summer but only the primary cooler is fired up all summer long. It's not until onion storage begins in mid-September that the second cooler sees heavy use.

Climate control equipment in these coolers is simple and inexpensive. Both coolers have window air conditioners and CoolBots for refrigeration in the summer and fall. Jonathan mentioned that the 24,000 BTUh air conditioner in the second cooler was undersized for the space and only capable of maintaining about 42°F (6°C). However, this suits its use as an overflow cooler during early harvesting and as a predominantly winter storage space. Jonathan mainly uses

Jonathan Stevens with stacks of onion sacks in the secondary cooler at Jonathan's Farm. *Photo courtesy of Jonathan Stevens.*

the second cooler for crops that prefer drier winter storage conditions, such as onions and squash. The primary cooler has two 12,000 BTUh air conditioners and is kept around 40°F (4°C) during the summer. This cooler houses the majority of the root crops that prefer higher humidity, such as carrots and beets. The target winter temperature for both coolers is between 33 and 35°F (0.6 and 1.7°C). During the winter, Jonathan adds supplemental heat via small resistance space heaters controlled by Inkbird thermostat controllers, especially in December, January, and February, when outdoor temperatures sometimes drop to −30°F (−34°C). He doesn't control the humidity in the coolers; instead, he relies upon woven-plastic sacks to create humid microenvironments for the crops that need it.

Jonathan described some of his ongoing experiences with condensation in the coolers. The fact that both coolers serve double duty as summer and winter storage spaces is leading to problems with condensation buildup in the walls and ceiling. Jonathan sees excessive condensation in the ceilings during the summer, with water sometimes pooling to fill light fixtures. The coolers' walls and ceilings have double vapor barriers—foam board and taped seams on the inside and outside of the framing—and the excessive condensation in the ceilings likely means that warm, humid air from outside is getting through leaks in the outer vapor barrier. Jonathan said this moisture problem only occurs during the summer, noting that relative humidity in the wintertime coolers is not very high.

Harvest and Curing

As is common with farms doing both summertime and wintertime sales, Jonathan's Farm must harvest storage crops while continuing harvests and deliveries for the summer CSA and markets. In some ways, the storage-crop harvest complements the summer CSA work, since the storage harvest fills the void as succession plantings and weeding drop off in the late summer. Even so, Jonathan purposely schedules a slowdown toward the end of the summer CSA in October, reducing the number of weekly deliveries from two to one. They rely on this extra time to get the storage crops out of the ground, washed, and into the coolers.

The crew harvests most crops by hand, so their harvest systems and procedures have to be efficient. Jonathan's Farm uses a bed system with 30-inch (76.2 cm) beds and pathways that match the tractor's 5-foot (1.5 m) wheel spacing. They're able to outfit the tractor with forks to carry either pallets or plastic macrobins anywhere in the field. The crew also drives pickup trucks directly in the field to load and haul harvested veggies; the truck tires mostly fit between the beds.

Even with the extra time for harvesting, they take advantage of good weather and sometimes need to make hard pushes to get certain crops out of the ground. Jonathan said this is especially true for potatoes and carrots. Because the farm's soils are heavy clay, excessive moisture, along with light freezes, can make harvesting these crops nightmarish, forming blocky chunks that are difficult to dig through and remove when washing. Jonathan said he'll sometimes hire a large crew of ten or more people for a single day to get carrots out of the ground when heavy rains are expected. Wet soils are less of a concern for beets, turnips, and celeriac because these crops sit more on top of the soil.

Onions are the first storage crop harvested each year and usually come out by mid-to-late August after the tops have fallen. Crews remove the majority of the tops, leaving roughly 3-inch (7.6-cm) stubs, while harvesting. Onions cure on tables, and sometimes the floor, in the nursery greenhouse for two to three weeks before the process of cleaning, bagging, and moving them into the coolers begins. Jonathan relies on two large end-wall fans to move fresh air inside to assist drying. He prefers that onions go into storage clean, and the crew does this task whenever there's time in the early fall. Cleaned onions go into poly-mesh bags and are stacked on pallets in the secondary cooler.

Winter squash are the next storage crop to come out, and Jonathan tries to harvest them before the first frosts arrive in mid-September. Winter squash also go into the nursery greenhouse for curing—the onions are usually out before winter squash need the space. The crew load squash directly into plastic macrobins in the field, and they use a tractor with forks to drive bins into the greenhouse. Jonathan used to spread squash out in a single layer on the tables and floors but now leaves them in the bins in an effort to reduce both handling and subsequent damage to the squash. After several weeks in the greenhouse, squash go into the second cooler with the onions. He mentioned ongoing problems storing winter squash and estimated losing roughly 20 percent each year, mostly to black rot.

The potato and carrot harvests occur simultaneously starting around mid-September. These are the two largest plantings of storage crops, and the crew chips away at harvest whenever the weather is dry and they have time. Jonathan prefers that carrots receive at least one frost before they're harvested but said that he can't always afford to wait. Before harvesting carrots, Jonathan runs an undercutter bar to make hand-harvesting easier and faster. Crews rip off carrot tops, leaving residues behind, and pile carrots into 50-pound (22.7 kg) harvest bins carried on a pickup truck. For potatoes, they use a one-row digger to bring spuds to the soil surface and collect them similarly to carrots. Both potatoes and carrots—and all other root crops that require washing—are washed before going into long-term storage. After washing, carrots go into woven-poly sacks and potatoes go into 50-pound mesh bags.

Beets, turnips, celeriac, kohlrabi, and cabbages are the last crops harvested—often after hard frosts arrive—because they're easy to harvest even if the ground is wet. Jonathan and the crew pull these crops by hand without undercutting. They rip off beet and turnip tops by hand but cut celeriac tops, leaving all residue behind in the field. Jonathan noted that they cut into the celeriac root masses to make washing easier, and they make straight cuts on the tops knowing that they'll likely need to be retrimmed after storage. Beets, turnips, celeriac, and kohlrabi all go into the same woven-poly bags as the carrots to maintain humid microenvironments in storage.

Washing and Storage

The humidity-loving vegetables are stored in woven polypropylene bags similar to those used for holding 50 pounds of grain seed. Jonathan purchases these from a company in Winnipeg, and he said he's able to reuse the bags with minimal washing. Crops that want drier storage conditions, such as potatoes and onions, go into plastic mesh sacks. Winter squash and cabbages are stored in open plastic macrobins from harvest onward to minimize handling between harvest, curing (for squash), and storage.

The crew washes vegetables on spray tables prior to storage. Prior to washing, dirty roots are stored in sealed Rubbermaid bins in one of the coolers. As cooler space becomes limited, a crew member is assigned washing duty and spends a day emptying bins, washing roots, and transferring them to the appropriate bags for storage. Jonathan feels that this system works well and moves quickly even without more specialized washing equipment such as a barrel washer. Crew members empty the dirty roots from the Rubbermaid bins into bread crates spread onto a table and spray them clean. These crates also make for quick transfer back to the cleaned Rubbermaid bins, which are emptied into poly bags once they reach 50 pounds. All roots, even those destined for the sealed woven-poly bags, are bagged wet. Jonathan suspects bags allow enough air exchange to prevent the excess moisture from causing problems in storage.

Onions drying in greenhouse at Jonathan's Farm. *Photo courtesy of Jonathan Stevens.*

194

When I asked about the process of transferring washed roots from the Rubbermaid bins into the 50-pound bags, Jonathan said that the crew has developed an effective system for a single person to accomplish the task. Using a second Rubbermaid bin as a pedestal, the crew member slips the lip of the bag between the two bins and, with the lip of the bag pinched in place, tips the contents of the upper bin into the open mouth of the bag.

Both coolers can be accessed inside the roughly 1,200-square-foot washing and packing area. Crew members load bags of washed or processed veggies onto carts to wheel directly into the coolers. Veggies in 50-pound bags get stacked on pallets as high as they'll go. Ideally, the coolers are packed to the gills by

A crew member sprays off potatoes prior to storage at Jonathan's Farm. *Photo courtesy of Jonathan Stevens.*

early November. Jonathan sometimes has to use the second, drier cooler to house excess root crops in addition to onions and squash, thereby increasing the humidity inside, which sometimes causes problems with the squash.

We talked a little about washing veggies in the fall versus the winter. Jonathan said that he could wash veggies during the winter but that it would be energy intensive. The water in the wash shed freezes during the winter and requires a few days of heating to thaw. Jonathan does heat the wash-pack shed occasionally to prep for CSA drop-offs, but this time is minimal because of the market-style CSA (see the "Marketing" section below). Water drainage is also a concern with regard to winter washing. Drains in the wash-pack shed are piped into a drainage ditch, and the pipes might require heat trace cables to remain ice-free.

Marketing

Jonathan's Farm markets most of their winter-stored produce through a market-style CSA that starts in November roughly two weeks after the summer CSA concludes. Instead of receiving prepacked boxes, customers pay for 100 pounds (45.4 kg) of vegetables up front and take whatever amounts they want during each of four monthly pickup opportunities. Most customers choose to spread their purchases evenly over the pickups, but they could, for example, take 100 pounds of carrots at the first pickup. Jonathan saves a lot of time by bringing bulk bins to CSA pickups rather than packing boxes. Jonathan's Farm also sells winter vegetables wholesale to a couple stores in Winnipeg to soak up extras and leftovers from the CSA.

I asked Jonathan about the challenges of planning a market-style CSA with storage crops. After all, with the fixed inventories of winter storage, it's difficult to make mid-season adjustments if customer demand doesn't match what's available. Jonathan said that he's learned through experience what vegetable amounts and ratios to bring and that it's all surprisingly predictable. He sells winter CSA shares based upon each year's harvest and expects each customer to take roughly 24 pounds of potatoes, 24 pounds of carrots, 17 pounds of onions, 9 pounds of squash, 9 pounds of cabbage, 9 pounds of beets, 3 pounds of celeriac, 2 pounds of kohlrabi, 1.5 pounds of turnips, and 1.5 pounds of radishes. He also expects the average customer to take veggies in equal amounts over the four pickups, which helps him plan how much to bring each month. He tries to bring extras of the staple crops—potatoes, carrots, and onions—and promises that those will be available at every pickup. He's occasionally made special deliveries to customers when a staple crop unexpectedly ran out at a pickup. Customers are told that the other, nonstaple crops may run out before the final pickup opportunity and informed of how many months to expect them; for example, Jonathan may tell customers to only expect winter squash for two to three months. Jonathan stressed that bringing more than he needs, inventory allowing, to each pickup is important. If he unexpectedly runs out of something, customers will take more of something else, which throws off the estimates from the beginning of the season.

Grading and culling at Jonathan's Farm usually happens alongside washing and unpacking veggies from storage. When preparing veggies for the CSA, they dump the bags onto a table to remove the occasional soft or rotten piece before moving the veggies into the clean Rubbermaid bins that they bring to pickups. Jonathan said that his CSA customers are very accepting of weirdly shaped vegetables, so he does little grading based on shape. He also noted that customers selecting their own veggies at the market-style pickups creates a natural grading system, as they often leave broken or weird pieces behind. Jonathan gives these discards away to employees or friends, and they're often used as feed for chickens and horses. Culls and whatever discards are left go into a compost pile on the farm.

Favorite Varieties

Jonathan wasn't married to any storage-crop varieties except Bolero carrots. He mentioned growing both Boro and Red Ace for red beets and Patterson for yellow storage onions, but admitted he didn't need to pay too much attention to those.

Advice for New Storage Farmers

Jonathan encouraged growers interested in wintertime vegetable storage to try it, noting that it's easier than people might think. That said, he suggested starting with a few crops at a small scale. He added that as long as you can keep your storage room slightly above freezing, you're well on your way to success. Because humidity can be maintained with containers, whether sealed bags or bins, it's not always necessary to humidify an entire room to store high-quality winter veggies.

Open Door Farm

Owners: Jillian and Ross Mickens
Location: Cedar Grove, North Carolina; about 25 miles northwest of Durham
Farm size: 40 acres total; roughly 4 of 5 acres in production are dedicated to storage crops
Amount stored annually: 60,000 to 70,000 pounds (27,216–31,751 kg)
Signature storage crops: Winter squash, potatoes, and sweet potatoes make up the lion's share; other crops include onions, beets, daikon radishes, rutabagas, and cabbages
Winter markets: One farmers market and two CSA partnerships, both mid-August through March

Summer markets: Occasional sales through CSA partners; fresh-market sales in early fall
Portion of annual sales from storage: Roughly 80 percent
Number of people working: Jillian full-time, Ross part-time, and several occasional part-time employees plus harvest help

It's unusual to find farms focused heavily on storage crops in the American South, where winters are relatively mild. So I was excited to learn about Open Door Farm in North Carolina, whose owners focus their business on winter storage and offseason sales. Many Southern growers avoid storage crops because harvesting and readying them for storage can be difficult in the high heat and humidity common to the region. (The exception is sweet potatoes, of which North Carolina grows more than any other US state.) Storage farming in the South presents challenges and seasonal timings wholly unfamiliar to me, so I was grateful to chat with Jillian about the storage practices that she and her husband Ross have been developing around the 36th parallel.

Journey to Wintertime Storage

Jillian and Ross didn't start as storage farmers. They were pushed toward the idea as a way to slow down the hectic pace of year-round farming. After starting Open Door Farm at an incubator farm in 2012, Jillian and Ross began farming on their own land in 2015. They focused on fresh-marketed produce, microgreens, and cut flowers, sold mainly through farmers markets and their online store. Jillian said that most small farmers in North Carolina automatically become four-season farmers. To Ross and Jillian, the pace of year-round weekly markets and the endless stream of succession plantings was unsustainable and, as they figured out, unnecessary. Jillian said she realized that

Jillian Mickens shows off some root crops in the fields at Open Door Farm. *Photo courtesy of Jillian Mickens.*

she could spend parts of the year away from the farmers market while making the same amount of money and giving themselves time to recoup.

Jillian and Ross faced some challenges changing their farm's focus to winter storage crops. First, it's unusual to grow crops for storage in North Carolina. They had to learn to grow and process crops like storage onions and winter squash during oppressively hot and humid weather, without advice from other growers. I need to emphasize that the heat is oppressive. This May, for example, an early heat wave brought scorching temps in the mid-90s °F (around 35°C) that blanched all their green onion tops white. (Most of my Alaskan friends and I would melt in that weather.) They also had to adjust to much larger plantings to meet their sales needs all winter long, which meant new methods of cultivation. They needed some new equipment for handling the large and heavy harvests, and new infrastructure for maintaining proper storage conditions through the winter.

Storage Spaces

Jillian and Ross had a friend on the incubator farm who turned old refrigerated shipping containers into affordable coolers. They followed their friend's example and chose this option as well. Such containers, used for refrigerated goods on cargo ships and freight trains, take relatively little modification to get them running as farm coolers. The containers had integrated refrigeration units, but Jillian and Ross removed these and installed CoolBot systems instead.

Jillian talked me through the process they used to convert the insulated containers into farm coolers. First, they installed level gravel pads for the containers near their wash-pack shed as stable foundations. The original refrigeration components were designed to be easily removable for replacement and repairs. Jillian and Ross removed the rectangular panels holding most of the refrigeration hardware from the inside back walls and had a friend cut holes in the steel for the air conditioner. That same friend also fabricated custom brackets to support the butt-ends of the ACs.

On the inside of the containers, they installed wiring and outlets, and they reinsulated the areas around the ACs. They also modified the doors a bit. Altogether, purchasing an 8 × 8 × 20-foot (2.4 × 2.4 × 6.1 m) container, shipping it to the farm, and purchasing and installing the cooling equipment cost less than $6,000 (in prepandemic times).

Since starting their winter-storage venture, Jillian and Ross have increased the number of container-coolers to three: one that's 20 feet (6.1 m) long and two that are 40 feet (12.2 m) long. These containers have cooling systems to match the needs of specific crops. One of the 40-foot containers has a single 24,000 BTUh air conditioner and is used for storing winter squash and sweet potatoes at 60°F (16°C). The other 40-foot container has two 24,000 BTUh air conditioners to cool the space to 40°F (4°C) for potatoes and onions. The 20-foot container has a single 24,000 BTUh AC to hold beets, daikons, and cabbages between 32 and 34°F (0 and 1°C) from November through March. To maintain high humidity in the potato and root crop coolers, as well as during the curing step for sweet potatoes, they use small misting humidifiers controlled by humidistat controllers. In the winter squash and sweet potato cooler, however, they use a dehumidifier to keep the relative humidity down near 65 percent after sweet potatoes cure.

Jillian said the shipping containers do a great job of mitigating the large temperature swings common to North Carolina winters. Even if temps swing between 70°F one day and 20°F the next (21 to −7°C), the containers easily maintain their set temperatures. Jillian specifically mentioned a night that got down to 7°F (−14°C) without any problems with freezing veggies.

Materials Handling

I asked Jillian about the ways they move and store crops around the farm. When harvesting, she said, they typically harvest directly into bulb crates and load them onto trailers, truck beds, or pallets carried on tractor forks. Most of their produce also sits open in bulb crates in storage, except for root crops like beets and daikons, which go into tightly woven poly bags—like those used as 50-pound grain bags—set into the crates for easier stacking. Jillian said they've tried perforated plastic bags for root crops in the past, but the restricted airflow caused excessive condensation and rot. Inside the coolers, they place a row of pallets along one wall and stack bulb crates onto those. They usually make stacks of crates running along the opposite wall until there's a human-sized path remaining.

Jillian said they've reached a scale where plastic macrobins could benefit their farm, but concerns about space and efficiency have made them hesitant to make the change. Standard macrobins (footprints of 40 by 48 inches) are tight fits for 8-foot-wide shipping containers, at least when using spaces efficiently and stacking two macrobins side by side. With their current system, they can nimbly grab any crates they need within the coolers, but using macrobins would necessitate lots of shuffling bins in and out to gain access to various crops. Jillian also worries about the ergonomics of working with macrobins. When preparing orders for CSA partners or markets, they usually don't use up a full macrobin's-worth of any given crop, and without a bin dumper, options for getting crops out of macrobins are awkward. Overall, Jillian thinks it's easier and faster to simply grab whatever quantity of standard crates of packed vegetables they need from the coolers and go.

Harvest and Processing

Northern growers rely on certain indicators of crop appearance to know when it's time to harvest, but those indicators aren't reliable in North Carolina's hot, humid weather. If Jillian and Ross tried to follow those signs, their crops would likely suffer disease or heat damage. They need to harvest sooner, and they're learning through experience to identify signs that their crops are ready for harvest or are in trouble.

The first storage crop harvested at Open Door Farm is potatoes in mid-June through mid-July. Jillian and Ross intentionally plant both early and late varieties to

Ross Mickens unloads crates of sweet potatoes from a tractor-mounted pallet into one of the farm's refrigerated shipping containers. *Photo courtesy of Jillian Mickens.*

spread out the harvest and subsequent processing. They usually test-dig tubers and watch for slight dieback in the leaves before harvesting, but extreme heat stress might lead them to harvest early to avoid misshapen or damaged tubers. They use a single-row potato digger borrowed through a local farm-tool cooperative. Tubers are left dirty, put into crates, and stored in the 40°F (4°C) cooler. Jillian said they haven't noticed any problems with skipping a curing step and speculated that the potatoes might partially cure in the ground before harvest due to the heat.

Onions are the next crop harvested on Open Door Farm, and Jillian and Ross are still working through the challenges of growing and processing onions for storage in their climate. In the past, they harvested whole plants in mid-July and spread them onto drying racks in their covered pack shed. I asked if they wait until tops fall, and Jillian said if the necks have even begun softening, it's too late. The heat and humidity of midsummer encourage all sorts of field rots, so Open Door Farm must harvest storage onions before the obvious signs of physiological maturity. Even under fans, Jillian feels that the 8-foot (2.4 m) stacks of drying racks on cinderblocks don't remove moisture quickly enough. For the future, they're considering mowing off the onion tops prior to harvest—like they do with their garlic crop—to remove excess moisture, improve airflow, and hasten drying in the racks. They're also considering buying an onion topping table to make onion cleaning faster and reduce the time onions spend outside the cooler. I also inquired if the high relative humidity in the cooler onions share with potatoes might contribute

Part-time helpers wash and sort rutabagas with the help of an AZS barrel washer and sorting table. *Photo courtesy of Jillian Mickens.*

to storage rots. Jillian wasn't sure but said the outer onion scales come out of the cooler soft and moist but quickly dry again. Ross has been researching onion storage techniques in other hot-climate regions and has found examples of hot-dry onion storage that they might try in the future.

Winter squash mature on Open Door Farm between mid-July and late August depending on the variety. As with potatoes, they intentionally plant varieties with a range of maturation times to make sure they're not overwhelmed by a flush of squash all at once. Like onions, winter squash cannot be left in the field until they show signs of maturity—corky stems, for example—commonly used by northern growers. As Jillian put it, "They'd be boiling hot snot balls if we let them get to that point." Instead, they check for other indicators of maturity like color changes on the ground spot, the sweetening and darkening of the flesh, and the size of seeds. Jillian noted that seed size has been a particularly reliable indicator of maturity—the squash are ready when seeds have filled out in girth. For over a month, they watch their acre of squash closely and harvest the fruits as soon as they're ready. Harvested squash sit in stacks of bulb crates in their pack shed for up to three weeks until the fruit can be processed. Although there's no dedicated curing step for winter squash at Open Door Farm, Jillian suspects the wait time in the warm pack shed might be filling this function. Before being moved into the 60°F (16°C) cooler with the sweet potatoes, winter squash are cleaned and sanitized in a rinse conveyor to reduce the pathogen load entering storage. They see few squash deteriorating early in storage, and their main losses are from squash reaching the ends of their storage lives.

In mid-September, it's sweet potato time! Jillian has a love-hate relationship with sweet potatoes. They're a great seller but can be difficult and tedious to harvest. Before running any harvesting equipment, they have to brush-hog the vines to clear a path for the diggers. In the past, Jillian and Ross undercut their sweet potatoes and dug them by hand, but they've been experimenting with harvesting equipment borrowed from friends. They've landed upon a Spedo-brand chain-style digger that harvests a single row. This mechanical digger leaves tubers on the soil surface and makes the harvest both easier and faster.

They harvest sweet potatoes into crates and cure them in the 20-foot (6.1 m) cooler at 85°F (29°C) and high humidity for five days. They're not able to harvest quickly enough to cure the entire crop at once and instead cure in batches. Once a batch has cured, they move it to the 60°F cooler with the winter squash.

The last storage crops harvested on Open Door Farm are beets, daikons, rutabagas, and cabbages, all of which are planted in August. These are usually harvested before mid-December, when there's bound to be a devastating cold event that kills most exposed crops. Jillian and Ross hand-harvest these crops but have been experimenting with ways to make the process more efficient and ergonomic. Lately, they've set up a pallet at working height on the tractor forks as a platform on which they can both top and grade the root crops in the field. They seal the graded and dirty crops into woven-poly bags, and set those into bulb crates stacked in the 20-foot (6.1 m) cooler.

Because of their warm climate, Open Door Farm has no trouble washing veggies all winter long. (This allows them the convenience of storing crops dirty.) They house all their washing equipment, including an AZS-brand barrel washer, rinse conveyor, and circular grading table, in their wash-pack shed. They use the barrel washer to clean beets, rutabagas, daikons, and potatoes, and they wash in batches of 120 to 150 pounds (54.4 to 68 kg) for eight to ten minutes. In addition to soil, the barrel washer removes sprouted eyes from potatoes nearing the end of their storage lives. Open Door Farm uses the rinse conveyor to wash both winter squash and sweet potatoes. Their rinse conveyor has three washing stages that they can customize depending on what they're washing. Jillian said the rinse conveyor is essential for the sensitive sweet potatoes, adding that without it, they couldn't grow as many as they do.

Bins of harvested squash in the packing shed await washing and sanitizing before storage. *Photo courtesy of Jillian Mickens.*

Marketing and Grading

Open Door Farm sells its produce from August through March. Some of those sales are of fresh-marketed crops, but stored veggies generate nearly 80 percent of their annual income. One major stream of income is the Saturday Carrboro Farmers' Market. Switching focus to storage crops has made them more competitive at market both in terms of timing and offerings. "Instead of struggling to sell tomatoes year-round, I can easily pull 100 pounds of potatoes from storage and sell them in two hours," Jillian said. Open Door Farm also makes about 30 percent of its annual income through partnerships with two local CSA farms, supplementing them with storage crops—a win-win relationship, as Jillian put it.

Most of the presale grading on Open Door Farm is based on size. This is particularly important for sweet potatoes, since customers often avoid jumbo-sized tubers at market. As previously mentioned, they grade root crops like beets and daikons in the field based on size to have a variety of offerings at market. Spoiled crops culled from storage go into a (mostly unmanaged) compost heap.

Favorite Varieties

Jillian always prefers varieties that last a long time in storage. A favorite potato variety is Huckleberry Gold. She said it stores well, tastes good, has an attractive look, and is an overall high yielder. She loves most butternut squash but said that Metro has been particularly robust in storage. Her customers love Sunshine kabochas, and it's her best-selling squash variety. Jillian also likes the kabocha/butternut hybrid Tetsukabuto, saying it lasts forever. She restated her love-hate relationship with sweet potatoes but mentioned Murasaki as an all-around solid variety.

Advice for New Storage Farmers

Jillian believes there is a lot of opportunity for farmers in North Carolina to expand into winter storage, but there are challenges to overcome as well. She thinks that farmers could benefit from the slower pace and the less-perishable crops, but she cautioned that to make money and justify equipment purchases with storage crops, you need larger plantings compared to more-valuable summer crops. Jillian was also enthusiastic about using retired shipping containers as coolers. She said they're perfect for new farmers because they're cheap, it's easy to get them functioning, and they will likely outlive most farmers' needs.

West Farm

Owners: Angus Baldwin and
Holly Simpson Baldwin
Location: Jeffersonville, Vermont; about forty-five
minutes northeast of Burlington
Farm size: 10 acres in vegetable production
with 7 acres dedicated to storage crops
Field setup: Permanent raised beds with spacings
matched to the tractor's wheel-width,
about 5.5 feet on-center (1.8 m)
Amount stored annually: 100,000 pounds
(45,359 kg)
Signature storage crops: Carrots, potatoes,
onions, cabbages, garlic, winter squash, and sweet
potatoes, plus smaller quantities of a few others,
including sunchokes
Winter markets: Mostly wholesale but with an
80-member winter CSA
Summer markets: Summer and fall CSAs alongside
sales to their wholesale distributors
Portion of annual sales from storage:
Roughly 60 to 70 percent
Number of people working: Angus full-time plus
six full-time crew members in the summer and one
part-time employee to help pack winter orders

There are several reasons why I was excited to interview Angus Baldwin and Holly Simpson Baldwin of West Farm. Partly, I was inspired by the farm's evolution from a small plot of land on an incubator farm to a thriving, multi-pronged business. Over the years, Angus and Holly identified opportunities and intentionally expanded in those directions. Early on, they recognized storage crops as a way to expand the business and fill a hole in their local food system. The couple also identified aspects of the farm that brought them fulfillment. For Holly, it was fermentation, and she eventually started a fermentation business as an offshoot of the farm.

West Farm expanded its scope and its winter storage in phases. Holly and Angus started Three Crows Farm in 2012 on land owned by the Vermont Land Trust, just down the road from their current location. Between 2012 and 2016, they grew from ¾ acre to nearly 3 acres, and during that time they started storing winter crops in their walk-in cooler. When an opportunity came to move onto a former dairy farm with more land and a large, well-built dairy barn, they took it. It took a few years, but they slowly converted the dairy barn into a superbly functional space for processing and storing both summer and winter crops. The additional land has allowed them to expand into wholesale markets in addition to their CSAs. Holly still runs Three Crows Farm as a value-added fermentation business, and the Vermont Land Trust still collaborates with West Farm to provide outreach and education opportunities to the community.

Storage Spaces

The legacy of dairy farming on West Farm's property meant that Angus and Holly could adapt existing infrastructure to the new farm's needs. The stanchion dairy barn, measuring roughly 35 × 140 feet (10.7 × 42.7 m), was structurally sound but needed improvements to make it functional for veggie storage and processing. One limitation was that the entire building had 8-foot (2.4 m) ceilings, and Angus would have preferred more vertical space. As soon as Angus and Holly moved to the property in 2016, they and their crew began renovating the barn.

During that first year, they built an 18 × 20-foot (5.5 × 6.1 m) cooler on the end of the barn with an exterior loading dock. They framed in two walls and sheathed the walls and ceiling with 4 inches of foil-faced foam board to prevent thermal bridging. Over that, they added wood fastening strips followed by sheet metal. This cooler is the primary cooler used

While renovating the second half of the barn at West Farm, Angus and Holly decided to add in-floor radiant heating to a section of the slab. *Photo courtesy of William Shinn.*

during the summer, in addition to its use for winter-time storage. The cooler has an 8-foot overhead door for access to the washing/packing space and loading dock. I'll refer to this cooler as the "primary cooler."

The remainder of the barn needed significant improvements to the concrete floors, and they began tackling the first section (about 30 feet, or 9.1 meters, of the barn's length), closest to the cooler, during the second year of operation. Angus said they did the work "farmer style," using reciprocating saws to remove stanchion poles and jackhammers to smash apart concrete curbs and add spaces for floor drains. Afterward, they hired a concrete contractor to refinish the surface and smooth out the mess they'd left behind. This area became the primary washing and packing area on the farm. They also cordoned off a small part of that space to store cured onions and garlic.

It took a couple years to secure the funds to renovate the remaining half of the barn. After receiving a grant in 2020, they used a skid steer mounted with a hydraulic concrete breaker to crack and remove the old slab of the final 75 feet (22.9 m) of the barn. Before pouring the new slab, they installed radiant in-floor heating tubes in a narrow section adjoining an exterior wall, and they split the in-floor heating into two zones that became separate rooms. One of these rooms is the warm-dry storage room that's used to cure and store sweet potatoes, as well as store winter squash during the winter. The other is the processing space for Holly's fermentation business. They split the remaining space (without in-floor heating) into two additional rooms: a winter-use cooler measuring roughly 40 × 23 feet (12.2 × 7.0 m) and an area for packing and for dry-processing crops like onions and garlic. Walls for the winter cooler and warm-dry storeroom were framed with 2 × 6 studs, insulated with fiberglass batts, and sheathed with OSB followed by sheet metal. Angus said that if he could do it again, he'd use rockwool instead of fiberglass, and he'd use wooden strips to support the sheet metal rather than OSB. Both the fiberglass and OSB have had issues with moisture, but it hasn't

become a serious problem because the spaces are only used seasonally.

Both the primary and winter coolers have commercial cooling units with remote evaporators and condensers. The condenser for the winter cooler is inside the pack room, partially heating that space during the winter. West Farm does not actively humidify either cooler during the winter and instead relies on storage containers to hold in moisture when necessary. The primary cooler—the one that's also used during the summer—stays more humid than the winter-only cooler and houses crops that require higher humidity. During the winter, Angus rewires the evaporator fans in the primary cooler to run only when the compressor is active. He said this can sometimes cause the evaporators to ice up (because they rely on that air circulation to defrost) but that it helps hold the overall humidity higher.

The winter-only cooler tends to stay drier because of its location in the middle of the barn, which Angus said is a flaw in the overall floorplan. To get from one side of the barn to the other, workers must pass through the winter cooler, which means frequent influxes of dry air. The humidity usually stays between 75 and 85 percent and makes this cooler well-suited for crops like potatoes and cabbages. The target winter storage temperature is around 38°F (3°C) in both coolers. Angus speculated that the ideal storage temp might be lower, because they sometimes see early sprouting, but he's been concerned about taking the temperature too low.

The target temperature for the squash and sweet potato room is 55°F (13°C). The choice to install radiant floor heating for sweet potatoes and squash was based on advice from another regional sweet-potato and winter-squash grower. In their experience, in-floor heating was more effective than convection heaters at keeping humidity down around the target of 60 to 70 percent. The allium room is kept between 38 and 42°F (3 and 6°C) and cooled with a CoolBot system in the fall and spring. For most of the winter, however, this space is cooled passively. Angus tries to keep the humidity there around 65 percent and said that controlling moisture is very important for good allium storage. He also stressed the importance of keeping the room cool in the spring when outdoor temperatures rise. That's when he engages the Cool-Bot to prevent late-season sprouting.

Materials Handling

West Farm uses a mix of bags, wooden bins, and plastic bins for storage. The first storage containers they used were woven-poly bags, similar to 40- or 50-pound seed bags. They hold in moisture and are relatively easy to move and stack. The farm's first large pallet bins came from a local sawmill and had planks for sides rather than plywood. The farm still uses these bins, but they try to avoid using them with potentially messy crops like winter squash. Eventually, West Farm purchased a set of 20-bushel plastic bins, which Angus said are the most durable and easiest to clean. Instead of bin liners, they pile a few full woven-poly bags on top of crops in the plastic bins to hold in moisture. Most crops that need high humidity go into the primary cooler, and any that must go into the drier winter-only cooler are stored in woven-poly bags.

To move crops from the field to the barn, West Farm usually outfits a tractor with forks. As the crew harvests, they either pile veggies directly into a pallet-sized bin or they hand-harvest into smaller plastic containers—either ⅝-bushel or 1-bushel—and shuttle those back to the larger bin. The barn does not have enough space to run a forklift inside, and the coolers especially can get too tight. The crew typically makes stacks of two pallet-sized bins outside the barn with the tractor. From there, they use pallet jacks to move bins to their final locations inside. Most orders are packed in the wash-pack area, which has easy access to an outdoor loading dock.

Harvesting and Processing

The West Farm crew follows a natural harvest order as crops mature, but they also pay attention to frost and pests like white-tailed deer. Beyond frost, they need to watch the weather for precipitation. Their soils are well-drained and rocky, but late-season rains can hamper harvesting activities.

Garlic is the first storage crop harvested at West Farm and usually comes out in late July. West Farm has been increasing the amount of garlic they grow, harvesting around 1,000 pounds in 2022. They try to pull garlic before the plants have dried too much to avoid split bulbs. They don't need to loosen the soil for hand harvesting—the crew just pulls the whole plants into bins. Back at the washing station, they spray the bulbs and roots clean and clip the tops down to 1- or 2-inch (2.5 or 5.1 cm) stubs. Cleaned and clipped garlic bulbs cure under fans on wire racks stacked on cinderblocks in the dry-processing area of the barn. Using this method, they can dry about 1,000 pounds (454 kg) of garlic in an 8 by 16-foot (2.4 by 4.9 m) area of floor space. Because they wash the bulbs at harvest time, the garlic takes little effort to spruce up later when packing orders. Angus mentioned that they clip stalks outside the field to avoid creating cumbersome piles. Instead, garlic stalks go into a compost pile after they're clipped.

Angus prefers to harvest onions after all the tops have fallen, but sometimes disease pressures such as downy mildew force them to pull onions earlier. West Farm grows all their onions on plastic mulch. At harvest time, they pull the onions and allow them to dry down on the plastic for a week or two before bringing them inside. I asked if rain was a concern, and Angus said that he's become more tolerant of occasional rainstorms dampening the drying onions. He used to fastidiously prevent onions from getting wet during the drying process, fearing that moisture would lead to storage rots. However, after talking with Quebec-based shallot growers and changing his own practices, Angus no longer thinks it's an issue. After the onions spend several weeks drying on the plastic, they're loaded, tops and all, into pallet bins that are stacked in piles of two in the dry-processing area of the barn. There, fans blow fresh air over the bins to complete the drying process. Angus said that filling bins with whole plants has saved both time and space. The practice keeps the onions relatively consolidated until the crew tops and cleans them later in the fall. After cleaning the onions by hand—a time-consuming process—they go into the allium cold room with the garlic.

Winter squash are typically the next crop harvested at West Farm and must come out before the first frost. The crew gently harvests squash into plastic pallet bins before stacking those in ambient conditions in the barn. Angus said that maintaining low humidity is important to preventing storage rots and that running fans over the squash helps bring humidity down. He also mentioned that they separate their squash based on expected storage life. The crew goes through the squash meant for short-term storage—mainly *C. pepo* types like delicatas and acorns—a few weeks after harvesting to remove bad pieces, packs up what they can sell soon, and puts the rest into bins for storage in the warm-dry storage room (after sweet potatoes have cured). The crew processes squash meant for longer-term storage—mainly butternuts and *C. maximas*—in a similar manner in January. Celeriac are one other crop that easily sustains frost damage, so the crew also tries to harvest these before frosts arrive.

Angus tries to get sweet potatoes out of the ground before soil temperatures drop below 55°F (13°C). West Farm grows sweet potatoes on plastic mulch, which they remove prior to harvest. Mowing the tops and removing the trailing vines makes removing the plastic mulch much easier. They used to undercut the sweet potatoes and dig them by hand, but now they use a single-row potato digger, which is faster and more effective. After harvest, sweet potatoes cure in the storage room with radiant floor heating set to 80 to 85°F (27 to 29°C) for five to seven days. Angus said it's possible to cure sweet potatoes in greenhouses but that it's preferable to do it in a dark room. After curing, they open the doors to reduce humidity and temperature until they can hold the room near 55°F. Angus said that properly cured sweet potatoes can last

Crew member at West Farm emptying bins of freshly washed garlic onto the drying racks in the dry-processing room. *Photo courtesy of Holly Simpson Baldwin.*

until June the following year, but he typically sells his by the end of April.

Although potatoes can stay in the ground until it starts to freeze, Angus prefers to harvest them after the sweet potatoes to limit the buildup of pathogens. In particular, he said he often sees more issues with *Rhizoctonia* fungi the longer potatoes stay in the ground. The farm purchased a single-row potato digger after a few years of digging ¼ acre of potatoes by hand, and Angus said it was money well spent. Digging the potatoes earlier also reduces the chances of dealing with wet soil, which makes the digger less effective. Angus noted that wherever he's run the potato digger, either for potatoes or sweet potatoes, he runs a subsoiler or ripper the following spring to prevent a hardpan from forming.

Carrots are another crop that Angus prefers to harvest sooner rather than later. Even though carrots are quite resilient to cold weather, the farm lies within a white-tailed deer migration corridor, and they love carrots. Angus said it's not unusual for the deer to eat 100 row-feet of carrots in a single night if they're left unprotected. The crew breaks off tops by hand and leaves them in the field to be tilled under the following spring. They harvest beets in a similar way, and crew members rip off beet tops just like the carrots.

Brassicas are hardy enough to delay harvesting until threatened by heavy frosts. The crew harvests cabbages into pallet bins before bringing them into the cooler. They like to leave on one to two wrapper leaves—which Angus described as leaves not entirely tight to the head—to protect the cabbages from drying out in storage. Good storage varieties easily last

The crew at West Farm harvests sweet potatoes after bringing the tubers to the surface with the potato harvester. *Photo courtesy of Holly Simpson Baldwin.*

into March without much loss. Brassica stalks are cut either with a brush hog or a flail mower in the spring.

West Farm stores all of its root crops dirty and uses either a barrel washer or a brush washer to clean crops as they're needed for sales. They use the popular AZS barrel washer purchased through Market Farm Implement (see resources, page 237), sending crops through in batches. A single person can operate the washer, which has a variable-speed motor to make adjustments during washing. For sweet potatoes, West Farm uses a brush washer, which tends to do a better job cleaning this crop. The farm doesn't have a bin flipper, and they use their plastic ⅝-bushel buckets to transfer crops from storage bins into the washer. After they're clean, veggies drop into cleaned bins, and the crew packs orders from those.

Sunchokes

Angus was the only farmer I interviewed currently growing sunchokes (a.k.a. Jerusalem artichokes) commercially. (I used to grow them at Root Cellar Farm, but they were more of a wind break than a dedicated crop.) For the effort he puts in, Angus thinks sunchokes are one of the most valuable storage crops at West Farm. That said, sunchokes are aggressive perennials that can take over an area if they aren't managed well. West Farm plants sunchokes in early June, putting down one row per bed (12 inches, or 30 centimeters, in-row spacing) with their water-wheel transplanter. After planting, the sunchokes don't require much maintenance other than light cultivation to thrive. The crew can harvest the tubers anytime in the late fall before the ground freezes, but they usually harvest them last. Angus said they've tried digging the sunchokes by hand but that the single-row potato digger is the most effective tool. He also noted that volunteers are inevitable the following year wherever sunchokes are planted, and those volunteers readily grow through plastic mulch. Angus uses a year-to-year rotation with sunchokes, harrowing the area planted the previous year to keep the volunteers from competing with other crops. Angus estimated

that they got about 1,000 pounds from 600 row-feet last year, which took an afternoon to harvest with a couple of people helping.

Marketing and Planning

From the beginning, one of West Farm's goals was to fill gaps in the local food system, and winter storage was an important part of achieving that. The bulk of their winter storage crops sell through two regional wholesale distributors. The farm also offers a winter CSA to its employees and the local community, filling about 80 shares annually. Angus said that the winter CSA is in higher demand than their summer CSA, noting that many other farms offer summer CSAs, whereas local veggies are sparse during the winter. Between all the marketing outlets, West Farm sells between $100,000 and $120,000 of produce between late October and mid-April.

One benefit of growing crops for a regional wholesale distributor is that West Farm is able to grow larger amounts of less popular crops, especially certain brassicas like storage radishes, turnips, rutabagas, and kohlrabis. Angus said the wholesale markets are usually able to absorb any excess crops if he's willing to drop the price. This allows him to be somewhat speculative when growing certain crops. Even if one potential wholesale buyer doesn't come through, Angus can usually sell a surplus through another.

A downside of selling through regional distributors is the stringent grading requirements. Angus said their grading has to be fairly strict, and the crew tries to exclude pieces that won't pass muster in the field to save on overall storage and processing space. Carrots, for example, must be straight, properly sized, and without major defects. He mentioned that forking can be a problem, especially with plantings seeded early in the summer.

West Farm packs the majority of wholesale orders into 25-pound polypropylene bags. Their winter CSA veggies go into corrugated boxes made from recycled plastic by an American company called CoolSeal USA (see resources, page 237). The veggies are mostly

Late-fall cabbage harvest at West Farm. These cabbages were destined for the storage barn in the background. *Photo courtesy of Holly Simpson Baldwin.*

Favorite Varieties

Angus likes the green cabbage variety Expect, which stores well and is the right size (3 to 4 pounds, or 1.4 to 1.8 kilograms) for wholesale markets. Avalon butternut squash have consistently stored into February, and the butternut-pumpkin cross Autumn Frost has done very well in storage, consistently outlasting other pure butternut varieties. Bolero, as usual, is the choicest carrot for quality and longevity in storage. Angus said he likes Boro red beets for their shape, size, and eating quality. He finds specialty and fingerling potatoes easier to market than table-stock varieties. Music is a popular garlic variety in Vermont, and Angus sees little reason to branch out to something new. He said Music is a good compromise between large cloves, which customers like, and low seed costs, since Music averages around five cloves per bulb.

Advice for New Storage Farmers

Angus advised farmers interested in winter crop storage to start small and gauge customer interest before starting. He noted the value of bagging products for local stores rather than selling loose, bulk veggies. Customers seem to prefer buying pre-weighed and bagged produce over loose vegetables, and bags often keep your produce looking better for longer. Stores don't always do the best job maintaining crop quality, and it reflects poorly on you if they let your roots go soft and floppy. He also noted that a bag with a sticker is a great way to brand your farm.

Angus also advised farmers to think through wintertime access and handling in and around their storage spaces. He speculated that many early storage farmers end up washing crops in rooms barely above freezing or in kitchen sinks, neither of which are enjoyable or efficient. During his first year storing crops, he had to walk a half-mile through snow to access the cooler—in the middle of the field—and bring veggies inside to wash in the kitchen sink. Neither prospect motivated him much, and he advised farmers new to winter storage to skip this torture entirely.

loose, with the exception of potatoes in paper bags, and customers bring their own containers to transport their veggies home.

Diseases and Pests

As previously mentioned, white-tailed deer can cause considerable crop damage at West Farm. Angus said the deer primarily go after chicory and carrots, but they sometimes eat peppers and have even found their way into the greenhouses. Thankfully, they do not typically eat brassicas on the farm. Temporary electric fencing is the main control strategy at West Farm, and Angus said they try to exploit the deer's poor depth perception when setting fencing. It's not infallible, but setting a single electric line in the wheel well surrounding the carrot beds often keeps deer out.

Tipi Produce

Owners: Beth Kazmar and Steve Pincus

Location: Evansville, Wisconsin; about half an hour south of Madison

Farm size: 40 acres in vegetable production with about 6 acres dedicated to storage crops

Field setup: Bed system spaced 6 feet (1.8 m) center-to-center to match tractor tires; 10-foot-wide (3 m) driving paths every eighth bed

Amount stored annually: 140,000 to 170,000 pounds (63,503 to 77,111 kg), depending on the year

Signature storage crops: Carrots are their most popular storage crop, but they also sell cabbages (including Napas), rutabagas, celeriac, radishes, leeks, parsnips, turnips, and beets.

Winter markets: Wholesale only between late November and April, except one CSA "Storage Box" around Thanksgiving

Summer markets: A 500+ member CSA and wholesale to stores in Milwaukee and Madison

Portion of annual sales from storage: 20 percent

Number of people working: Beth and Steve full-time plus twelve to fourteen full-time employees during the summer and six part-time employees during the winter

Tipi Produce is well known in southern Wisconsin as a leader in organic vegetable production. The farm is a great example of a summer-production powerhouse that recognized winter storage as a means of generating additional income and retaining core crew members through the winter lull. I found their journey into storage interesting because, unlike most other farms, the customers came to them asking for storage crops. The farm has been in its Evansville, Wisconsin, location since 2001, but co-owners Beth Kazmar and Steve Pincus have been farming in the area since 1999 and 1975, respectively. Because of their years of experience, smart conservation practices, and educational outreach, Tipi Produce was chosen as the Midwest Organic and Sustainable Education Service's (MOSES, now Marbleseed) Organic Farm of the Year in 2016.

Journey to Wintertime Storage

Tipi Produce first dabbled with winter vegetable storage in the mid-1990s. The Willy Street Co-op in Madison, Wisconsin, approached the farm asking for carrots longer into the winter. At the time, Steve thought it was a good way to increase annual income without any large infrastructural expenses. To cover the one large equipment expense—a Scott Viner harvester (see "Scott Viner Harvesters," page 26)—Tipi Produce received a loan from the Willy Street Co-op. At the time, the farm had one summer-use cooler that they recruited for winter storage. The cooler door was too narrow for a pallet jack to fit through, so they passed cabbages inside by hand, piling them loose in a corner. Carrots and other roots were kept in crates small enough to carry by hand. They washed produce in the greenhouse, which had water and heat during the winter. Eventually they purchased some used walk-in cooler panels and built another cooler specifically for winter storage. Instead of concrete, they laid a plywood floor that enabled them to move homemade 12-bushel bins with a pallet jack. This second cooler had no refrigeration until they moved to the current farm location. During those early years of winter storage, they relied upon cool weather to get the temperature in the winter cooler low enough for storing roots. Steve said this was sometimes a dicey proposition, and they were occasionally forced to shuffle bins to the cooler with refrigeration if the weather was too warm.

Storage Spaces

The farm uses five walk-in coolers and a refrigerated semi-trailer for storing winter produce, and three of those coolers are used during the summertime. The five walk-in coolers have commercial refrigeration units with remote evaporators and condensers, and the summer-use coolers have more cooling power than the winter-only coolers. The semi-trailer is refrigerated with the original diesel-powered cooling unit. Steve and Beth added insulation to the inside of the trailer to improve its energy efficiency, but Steve said it's still the most expensive storage space on the farm to operate. For that reason, they empty the semi-trailer first when they pull veggies from storage.

Tipi Produce maintains their coolers between 31 and 32°F (−1 and 0°C) through the winter, and Steve said that more cooling is needed than heating. In fact, there are some winters when they don't heat the coolers at all. "It's amazing how much heat a cooler full of carrots will generate on their own," he said. If the weather brings −10°F (−23°C) at night and temps just above 0°F (−18°C) during the day, then supplemental heat becomes necessary. To heat the coolers, they use resistance space heaters with thermostat controllers. Crew members turn off cooler condensers—with the evaporator fans left running for air circulation—whenever heating is necessary to prevent the cooling and heating systems from fighting with one another. The farm doesn't actively humidify the coolers because it's difficult to supply them with water in the winter. (There's only one winter water source, and it's not near any of the coolers.)

Tipi Produce also uses off-farm cold storage when they run out of storage space on the farm. Southern Wisconsin is a bastion of the cheese industry and has

Co-owner Steve Pincus forklifts bins into stacks near several of the coolers at Tipi Produce. The upside-down bin liners help hold in moisture. *Photo courtesy of Beth Kazmar.*

many refrigerated cheese warehouses that rent floor space to other food producers. It depends on the year, but Steve said they might send up to sixty pallet bins to a warehouse 30 miles away. The monthly rental rates are relatively low, but between trucking costs and warehouse access fees, it's pricey to move the veggies. While the warehouse temperature is around 35°F (2°C), the humidity inside is very low, and Steve said they must take special precautions to prevent roots from drying out.

Materials Handling

Tipi Produce uses mostly homemade storage bins. They hired a company to make skids, which are like pallets but lack the bottom decking that contacts the ground. The skids are standard pallet size (40 by 48 inches), have 4 × 4 lumber for outer runners, one 2 × 4 in the center for support, and top decking made from green, uncured oak which shrunk as it dried, leaving small gaps between the boards. On the farm, they used plywood to make 2-foot sidewalls and 2 × 6's for vertical outer supports. Steve said they tried ½-inch plywood but switched to ¾-inch because the thicker sheets made for stronger, more stable bins. (These weaker bins made with ½-inch plywood now only store cabbages.) After some experimentation, they realized that bolting the six vertical supports through the 4 × 4 runners was more stable than simply screwing them in place. These functional bins are stackable and continue to hold up after fifteen to twenty years of use.

They use a combination of forklifts, pallet jackets, and hydraulic pallet stackers to move bins around the

Pallet bins full of carrots sit on a harvest wagon at Tipi Produce. The bins were built on the farm using skids, plywood, and 2 × 6 supports. *Photo courtesy of Beth Kazmar.*

farm. Steve noted that the yard where the coolers sit is poured concrete, a legacy from the farm's past as a dairy operation. He said this yard is both stable and flat, and as long as they keep up with snow removal, it makes for an easily navigable surface in the winter. They can but rarely drive forklifts inside the coolers and instead maneuver bins inside with pallet jacks.

Harvesting and Processing

Each fall, Tipi Produce faces the challenge of harvesting large amounts of storage crops while simultaneously maintaining both their CSA and wholesale deliveries. Steve said that late October through mid-November is their crunch time. As such, the farm takes a break from CSA deliveries in late October to let the crew focus entirely on harvesting storage crops.

Tipi Produce uses specialized harvesting equipment to make the process fast and efficient. A key piece of equipment is a Belgium-made DeWulf harvester capable of harvesting carrots, beets, parsnips, and even turnips. Steve and Beth bought it secondhand from a dealer in the Netherlands. According to Steve, a three-person crew can harvest 3,500 pounds (1,588 kg) of carrots per hour in good conditions with this machine. The harvester does not perform well in wet soils because it has trouble staying aligned to the rows. Luckily, their sandy loam soils drain quickly after rain, and they're usually able to get back into the fields within a couple days of a hard autumn rainstorm. They also purchased a vibrating undercutter that they use to loosen soil beneath leeks, celeriac, daikon radishes, and rutabagas. Another key piece of equipment is a hydraulically operated harvest belt for cabbages that extends about 20 feet (6.1 m) into the field from a harvest wagon, minimizing the distance crew members have to travel while moving heads into bins.

All trimming occurs in the field during harvest, with residues left behind to be incorporated the following spring. Steve stressed that sharp knives are important for efficiency when hand-harvesting and trimming crops. For celeriac, crew members trim tops to about 1 inch (2.5 cm) and cut off the tangle of roots and soil without cutting into the meat of the root. For rutabagas, workers trim off the root mass on the taproot in addition to the tops. Crew members use lettuce knives—with squared or rounded tips—when harvesting cabbages so they can ergonomically push the knives through the cabbage stems rather than trying to cut sideways. They plan to retrim cabbages coming out of storage, so they leave three to four extra leaves at harvest time. For carrots, beets, parsnips, and turnips, the DeWulf harvester allows for adjustments to the topping mechanism, but Steve said it doesn't always cut entirely uniformly, sometimes leaving too much top or cutting into the root shoulders. This inconsistency hasn't been a problem, however. Steve said that any excess leaves usually come off during washing and that their customers don't mind roots missing a bit off the crowns. He also mentioned that the quality of the tops at harvest time affects how the harvester trims; high-quality tops lead to more consistent trimming.

Tipi Produce does not wash crops prior to storage, and storage containers vary by crop. Bins for cabbages receive a plastic liner that's left open during storage. Steve noted that in addition to special bins, cabbages have a special cooler that's slightly less humid than the others, which discourages mold growth in storage. Root crops stored on the farm go into the stronger wooden pallet bins without bin liners, but each bin gets a liner placed upside down over the top before going into a cooler. For roots headed to off-farm storage—mainly carrots—they place a liner inside each bin prior to loading it with dirty roots. The crew closes these liners with large twist ties before shipping them off. I asked Steve if they see problems with decay in the sealed bags; he said sometimes but mostly not. The alternative, he said, is bins of floppy carrots because of the low humidity in the cheese warehouse.

At Tipi Produce, they try to cool crops below 40°F (4°C) as quickly as possible after harvest because, in Steve's experience, this practice leads to fewer issues with bacteria or fungi in storage. If they're harvesting roots on a sunny day, they'll sometimes place shipping

Bins of carrots and parsnips fresh from the fields at Tipi Produce. These bins, with liners on the insides, are headed to off-farm storage where the liners will be sealed to hold in moisture. *Photo courtesy of Beth Kazmar.*

blankets over full bins to prevent them from baking in the sun. Temperatures within the coolers vary while loading in crops, but the goal is to keep cooler temps below 40°F. Sometimes this means slowing down the harvest to avoid overloading their refrigeration equipment. The crew also shuffles crops between coolers, using the three more powerful coolers to remove field heat before transferring bins to one of the winter-only coolers.

Washing and Packing Orders

Tipi Produce has an indoor washing setup that works all winter long; years ago, they walled off and insulated a portion of the barn to serve this purpose.

Steve said they were fortunate to have a relatively new dairy barn with high-quality concrete floors, 9-foot ceilings, and water hookups. An electric heater and thermostat hold the wash-pack area at 60°F (16°C) when the crew is working, but otherwise they keep the space at 40°F. There, they deploy either of two root barrel washers depending on the crop. They use an AZS barrel washer, which is gentler, for washing crops like yellow carrots, celeriac, and turnips in smaller quantities, whereas a larger, 8-foot barrel washer handles crops like orange carrots, beets, and parsnips in larger batches. To load the washers, they tip pallet bins onto the concrete floor with their hydraulic pallet lifter and shovel

Carrots exiting the large barrel washer are graded and packed into plastic totes for bulk wholesale during winter months. *Photo courtesy of Beth Kazmar.*

veggies into the washers with grain shovels. Crew members sort the veggies on a grading table after they exit the washer, separating them into different boxes depending on the grade.

Tipi Produce sells all crops except carrots in bulk and packs them into waxed or plastic boxes. The carrots sold in bulk are the highest quality, and they only put the straightest, most attractive carrots into this grade. They sell about two-thirds of their carrots in 5-pound, nonperforated bags. Their bagging grade is less stringent and includes double-forked, partially broken, and short carrots. Their lowest grade for carrots is juicing grade, which includes badly broken, small, and highly forked roots. Steve said these go to companies making retail juice, as opposed to juicing

bars, which typically want the largest and straightest carrots—"real honkers," as Steve put it.

After washing, they prefer to let vegetables dry overnight in a cooler before sealing the boxes or weighing carrots into bags. They usually do all the washing for orders in one day and all the packing the following day. Steve did admit that it doesn't always go to plan, and sometimes they need to put wet carrots in bags to quickly fill an order. That said, they've never received a complaint about their carrots deteriorating too quickly. Steve speculated that their carrots sell fast enough so that moisture in the bag doesn't become an issue. He also noted that their veggies are always kept cool, and they have refrigerated trucks for deliveries to Madison and Milwaukee.

Sales and Marketing

Tipi Produce has become a valuable partner for retailers in southern Wisconsin wishing to carry local vegetables through the winter. Despite recently reducing their summer wholesale in favor of more CSA sales, Steve and Beth see value in continuing their winter storage and wholesale program. Two large co-ops in Madison and Milwaukee account for about 85 percent of Tipi's wholesale sales, with a small restaurant distributor and a juice company comprising the remainder. However, the first recipients of storage crops are the CSA members. The last hurrah for the summer CSA program is a large storage box delivered just before Thanksgiving and containing the winter wholesale lineup in addition to garlic, onions, and winter squash. After Thanksgiving, Tipi Produce becomes a wholesale-only farm until about mid-April. They make deliveries to the Madison stores twice a week and once a week to Milwaukee throughout the winter. Orange carrots make up the majority of wintertime sales. When I asked about labels and specialty bags, Steve said they're still using their original method for labeling: small paper slips placed face-up inside the clear poly bags. Fall and winter storage cabbage is their second-biggest seller through wholesale outlets. Steve noted that they do not grow potatoes on their farm and choose to purchase potatoes for the summer CSA from other area farms.

Diseases and Pests

Tipi Produce uses crop rotations to limit the buildup and impact of diseases; they let at least four years pass before they grow any given crop in the same location. In the past, black rot—caused by the bacteria in the *Xanthomonas* genus—has severely affected brassicas like cabbages, rutabagas, and daikon radishes. They've managed this issue through hot-water seed treatment and crop rotation, so crop residue removal has never become necessary.

Steve noted that they occasionally see bacterial soft rot—usually caused by mixes of species from *Erwinia* and *Psuedomonas*—in stored carrots and beets.[1] This problem usually occurs in bins sent to the off-farm refrigerated warehouse. Because Steve and Beth keep track of which bins come from which fields, they're able to see that quality differs depending on where a crop is stored. Carrots from the same field might do fine in coolers on the farm, while those sent to off-farm storage sometimes "liquify," as Steve put it. He and Beth suspect that fully sealing the liner bags sent to off-farm storage encourages bacterial growth and crop decay. This hasn't been enough of a problem for them to make changes to how and where they store their crops, however.

All crops that go bad in storage or are otherwise unsalable go into a large refuse pile that they occasionally spread into grassy areas away from the fields. Steve said that much of what goes into the pile is diseased, and he doesn't want to bring those pathogens back into the fields. He would feel differently if they hot-composted the pile, but they don't make or manage compost on the farm.

The large pile of vegetable refuse, in addition to the farm's legacy of dairy, attracts a lot of rodents. Steve said that mice and rats have damaged stored crops, especially cabbages, in years when the rodent population has gotten away from them. They've tried active control methods including snap traps, live traps, and poison, but Steve said the most effective rodent controllers are cats. They recently added metal flashing to the bottom wall sections of several coolers to prevent rats from getting in, but the results of this change are yet to be seen.

Favorite Varieties

As a large CSA farm, Tipi Produce has experimented with many cultivars throughout the years. For storage carrots, Bolero is their consistent favorite. This variety is cold-hardy, tastes great, keeps extraordinarily well in storage, and doesn't get leaf diseases. Purple Top White Globe turnips have proved better than newer turnip hybrids on their farm. For radishes, Steve said they've switched over entirely to Korean-style daikons for their taste and customers' preference for their shorter length. Storage No. 4 has proved excellent as a green storage cabbage, but Steve prefers Capture for short-term storage. It's the most black-rot resistant cabbage variety he's come across. Ruby Perfection is a favorite for red cabbage, but Steve also mentioned Red Dynasty as a variety that's shown promise. Most varieties of Napa cabbage do well in storage until about February if they go in healthy.

Advice for New Storage Farmers

Steve advised anyone interested in winter crop storage to start small. Early in our conversation, he noted that a farm with a summer-use walk-in cooler can begin storing winter vegetables with little to no added costs besides utilities. He also suggested cabbage as a great starter crop because it takes little to no specialized equipment to harvest and process efficiently. Additionally, cabbage doesn't require any washing setup, wintertime or otherwise. He noted that farmers anywhere should have no trouble selling carrots and cabbages during the winter, but carrots take more thought, planning, and effort.

Food Farm

Owners: Janaki Fisher-Merritt and Annie Dugan

Location: Wrenshall, Minnesota; about thirty minutes south of Duluth

Farm size: 37 tillable acres, with 17 acres in vegetable production and 7 acres dedicated to storage crops

Field setup: Bed system spaced 6 feet (1.8 m) center-to-center to match tractor tires

Amount stored annually: 200,000 pounds (90,718 kg)

Signature storage crops: Carrots are the biggest seller, but they also grow garlic, onions, winter squash, beets, daikon radishes, parsnips, cabbages, and rutabagas

Winter markets: Both CSA (275 boxes per month) and wholesale (twice per week)

Summer markets: Traditional summer CSA, specialized shares such as garlic and basil, and wholesale deliveries (twice per week)

Portion of annual sales from storage: Roughly 50 percent by weight

Number of people working: Janaki full-time and eight full-time employees during the summer, and three to four part-time employees during the winter

When I was getting started with storage farming, Food Farm was a key source of both inspiration and

The storage building at Food Farm. The original building, built in 2000, includes a living space above the storage and processing spaces in the walkout basement. They expanded in 2015 with an attached backfilled structure, nearly tripling the total floor space. *Courtesy of Janaki Fisher-Merritt.*

information. I saw Janaki give a talk about winter storage at the MOSES (now Marbleseed) conference while planning my first season at Root Cellar Farm. I remember being impressed that a well-established multi-generational farm, even after thirty-plus years in operation, repeatedly increased its winter storage capacity. Janaki said it made sense for the local climate and there was huge customer demand for vegetables in the winter. I also remember feeling intimidated by the scale and cost of Food Farm's storage infrastructure. That said, I think Food Farm's history shows that storage crops are worth growing at many scales and, if the farmer wants to expand, storage crops can pay for the upgrades necessary to scale up.

Journey to Wintertime Storage

Food Farm was founded by John and Jane Fisher-Merritt as a vegetable farm in 1975 and is now run by their son Janaki. In the mid-1990s, Food Farm began keeping a small amount of storage crops beyond their regular summer and fall sales. The idea was to increase annual sales by selling more vegetables to the same set of enthusiastic customers, providing them with year-round access to local veggies.

When first dabbling with winter storage, they used a 5 × 9-foot (1.5 × 2.7 m) "fruit room" in their home's basement that held about 3,000 pounds (1,361 kg) of crops when full. After five years of using this as their primary storage space, the Fisher-Merritts built a dedicated root cellar and packing area capable of holding nearly 80,000 (36,387) pounds in 2000. Seeing that storage crops were profitable and the demand from the community remained unmet, they began planning to further increase their storage-crop production and build a new, even larger root cellar, which they finished in 2015. They built the facility attached to their existing storage and packing space and incorporated ideas from nearly twenty years of experience into the design. To pay for these expensive upgrades, Food Farm saved money, launched a donation campaign among their CSA members, and received a private loan from a customer.

At this larger scale, Food Farm also needed new storage bins and equipment for harvesting, handling, and washing crops. All in all, the project cost about $250,000 for the building, washing setup, storage bins, and handling equipment, but the farm started seeing dividends in their first season with the new facility. Looking back, Janaki said he has no regrets about the upgrades.

Storage Space Designs

The Fisher-Merritts built their first root cellar like a two-story home with a walkout basement. The basement level is the root cellar and packing area, while the second story is a living space. The basement is earthed in on three sides, while the open side has an overhead door for vehicle access to the storage area. It made both economic and practical sense to build a living space over the root cellar. Foundations and roofs are both expensive, so why build two when you can get away with one? Janaki, and eventually his family, moved into the older farmhouse when John and Jane moved above the finished root cellar.

When they built the new root cellar in 2015, it made sense to attach it to the original cellar. That way, the new root cellar, washing area, and packing line would add to their existing spaces, making one large indoor area to efficiently store and process veggies. The original building is roughly 24 by 52 feet, with about 1,200 square feet for storage and processing. The expansion added roughly 2,100 square feet of indoor storage and processing space and a 600-square-foot outdoor covered washing area.

The new root cellar and processing areas were designed to be functional and energy-efficient. The new building has 12-foot poured-concrete walls that were backfilled with earth to around 10 feet to take advantage of natural ground heat. They insulated these concrete walls both above and below ground level. Down to about 6 feet below the surface, the walls are insulated with 6 inches of XPS foam board. At the bottom of this foam, 2-inch XPS wings extend outward 8 feet to prevent frost penetration. Above ground level, they covered the exterior concrete walls with 10-inch structurally insulated panels (SIPs), which, in addition to insulating the walls, provided a means to easily fasten the sheet metal siding. SIPs (12-inch) were also used to create the peaked roof, with a central I-beam supporting the panels at a height of 18 feet in the center. Both the indoor and outdoor wash areas have 12-foot floor drains to collect and redirect water during washing. Concrete surfaces were left exposed on the building's interior, whereas any framing or SIP surfaces were sheathed with sheet metal as a water-resistant covering.

During construction, they insulated the concrete slab depending on the needs for each part of the new building. The slab was not insulated in the potato storage room since ambient soil helps keep the temperature in this space within a range suitable for potatoes, 38 to 40°F (3 to 4°C). The 32°F cold-storage room, on the other hand, needed insulation under the slab to prevent the ground from adding too much heat. Janaki feared the cold room might need supplemental heat during the coldest parts of the winter, so he left about a quarter of the room's slab uninsulated. Now, years later, he concludes he should have insulated the entire slab. Because the building is superbly insulated and the tons of respiring vegetables create a lot of heat, the extra ground heat just isn't necessary. In fact, the refrigeration system remains active throughout the entire winter to keep the room at 32°F.

The slab under the new wash-pack area received 2 inches of foam insulation. Lastly, the short concrete stem walls that support the interior walls were insulated with 2 inches of foam to prevent unwanted thermal bridging between different temperature zones.

Storage Spaces and Climate

The farm has three separate storage rooms with different conditions in each: a cold-storage room kept at 32°F (0°C), a potato storage room kept at 38°F (3.3°C), and a winter squash room kept at 50°F (10°C).

Food Farm keeps roots and cabbages in the cold-storage room throughout the winter. It's accessible from the main wash-pack area through an 8-foot sliding door and is roughly 20 by 36 feet with 12- to 18-foot ceilings. Two powerful commercial cooling units designed for industrial freezers hold the temperature as close to 32°F as possible, and these large evaporators help maintain high humidity in the cold room (see "The Basics of Cooling Equipment," page 109). Janaki mentioned an added bonus of having two cooling units: When one breaks down and needs service, the cooler can still function. Food Farm does not actively humidify the space, but the humidity stays around 80 to 90 percent throughout the winter. Janaki said that when the relative humidity approaches 95 percent, condensation appears sporadically on the walls and ceilings, which can lead to dripping and mold growth.

The adjacent potato storage room is a 16-by-36 foot space accessible from the cold-storage room through another 8-foot sliding door. They keep this room at 38°F through the winter but at a lower humidity than the cold-storage room. The lower humidity is a result of the cooling system, which relies on a differential thermostat that controls an air-intake fan. This thermostat triggers the fan to pull in cool and dry outside air when the potato room gets too warm. It also allows users to set a temperature differential—the difference between indoor and outdoor temperature—to control when the fan engages and indirectly control humidity (see "Humidity Control," page 119).

The older root cellar now acts as a storage space for onions, garlic, and winter squash, in addition to empty storage bins and other packing equipment. The farm stores winter squash in a portion of the old root cellar that's been walled off and insulated. A small heater and fan keep the room near 50°F and the air well circulated. The air temperature outside the squash room, which is open to the rest of the washing/packing area, varies between the 50s °F (about 10 to 15°C) in November to the low to mid-40s °F (about 4 to 8°C) by December and onward. Although these aren't the ideal storage temperatures for alliums, this

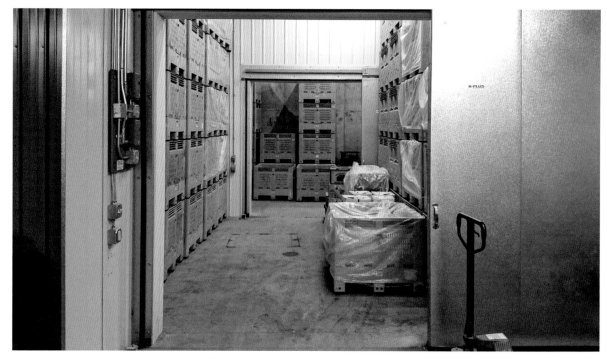

A view of both cold-storage rooms on Food Farm: the cold room, kept at 32°F (front), and the potato storage room, kept at 38°F (back). Notice the concrete wall in the potato room, which is mostly uninsulated from the soil beyond.

area is much drier than either the cold room or the potato room. Janaki explained that low humidity is more important than temperature to maintaining onion quality in storage.

Materials Handling

Food Farm has two sets of pallet bins: an older, home-made wooden set and a plastic set of macrobins purchased when they built the new storage area. Storage bins get scrubbed and sanitized each fall, and sometimes receive bin liners and/or burlap depending on the crop and whether it's been washed.

Food Farm uses a reach truck to move and stack bins within the storage and washing building. The reach truck is a small forklift that the operator rides standing up, and it has a tighter turning radius than a traditional forklift, making it great for tightly packed storage spaces. Using this, Food Farm can stack their storage bins five-high in the cold and potato rooms. Outside, they use a skid steer with forks to move bins

from place to place. They use buckets when moving crops by hand because they're sturdy, inexpensive, and limit the weight that workers can lift at one time.

Harvesting and Processing

Food Farm has a lot to harvest each year, and they carefully watch the weather to ensure the mechanical harvesters will work effectively and to harvest before cold weather causes crop damage.

The Food Farm crew hand-harvests and processes garlic with some assistance from tractor-mounted implements. When the garlic is fully mature, usually in early August, the crew trims stems down to 6-inch (15.2 cm) stubs with a three-point flail mower. After this, they loosen the soil by running a keyline plow—a type of subsoiler—between the rows. The crew then harvests, washes (lightly spraying dirt clods from the roots), and cures the plants on dog-kennel panels stacked in the production greenhouse. They cover that portion of the greenhouse with shade cloth to

222

prevent sun scalding. After several weeks of curing, the crew loads the dried garlic onto stackable bread crates and stores them in their heated machine shed. If necessary, they'll bring a dehumidifier into the shed to avoid mildew/mold growth on the garlic bulbs.

Onions are one of the few crops at Food Farm grown on plastic mulch, and when the tops begin to fall in late August / early September, the crew pulls and lays the entire crop on the plastic mulch to sun-dry for several weeks if the weather allows. Once the onions are dry, the crew tops them in the field and finishes the curing process in the same manner as the garlic. Onions remain in the greenhouse until temperatures threaten to stay below freezing. Janaki said they use onion cleaning as a rainy-day activity to keep the crew busy. Their hope is that by late October, all the onions are cleaned and ready to move to their final destination in the old root cellar.

The first forecasted frosts, which usually come in mid-September, are the signal to harvest winter squash. The squash can handle light frosts if they have a dense leaf canopy, but Janaki noted that late powdery mildew usually makes this impossible. Food Farm does not field-cure squash, but they try to clip kabocha-type squash (*C. maxima*) a few days before harvest to allow their thick, moisture-laden stems to begin drying. After harvest, the crew transports squash in shallow bins and racks to cure in the heated shop. They open the shop door during the day to let humidity out and fresh air in. Because the shop is easy to heat, they can postpone moving winter squash to the root cellar until they have time.

Food Farm grows a lot of cabbage, and the cabbage harvest can take upward of two weeks. Janaki said that they try to harvest cabbages and other crops with minimal leaf cover, like beets and radishes, before prolonged periods of below-freezing weather or any temps below 26°F (−3°C). (He noted that the outer cabbage leaves can be peeled away if frozen, but it's tedious.) The crew hand-harvests cabbages directly into bulk storage bins on a hay wagon driven into the field. They trim cabbages down to two outer wrapper leaves, which Janaki described as the innermost two leaves with tips curling away from the head. Janaki explained that these leaves protect the inner head during storage and are easy to remove quickly when packing cabbages into CSA boxes. The cabbages go directly into the cold-storage room and are stored without bin liners, which limit airflow and can cause the cabbage leaves to deteriorate and become slimy.

Janaki aims to harvest potatoes before cold temperatures damage the tubers, which can depend on both the weather and the integrity of potato mounds in the fall. Early cold snaps can sometimes cause freeze damage in the potatoes, even if mounds are relatively intact. A few years ago, Food Farm upgraded their potato harvester from a simple tractor-mounted single-row digger to an implement made by the Finnish company Juko. This new harvester digs potatoes, removes soil and residue, and deposits spuds into waiting bins. Now potatoes take only a few days to harvest with a small crew. Janaki noted that the harvester works best when the soil is dry. After harvest, potatoes go directly into the potato storage room and are stored either dirty or clean, without bin liners.

Food Farm grows roughly 2 acres of carrots and can complete the harvest in about three days. Janaki prefers to harvest carrots after at least one period of cold weather. Because full carrot bins often sit in the field for several hours before going into the cooler, it's also important to harvest when temperatures are cool. Hours of exposure to warm weather can dry out the carrot skins—making washing more difficult—can cause early decay in storage, and make the coolers work harder to remove field heat.

Food Farm uses a specialized harvester for carrots: the popular Scott Viner. Like the potato harvester, it can be difficult to keep the Scott Viner aligned on the rows when the soil is wet. This model harvests a single row at a time, topping roots and removing some soil before tossing them out a back chute where a crew member catches carrots into buckets. That crew member fills plastic-lined bins stacked on a modified boat trailer pulled behind the harvester. The bin liners are extra-long so the crew can fold over the excess to hold in moisture and prevent desiccation in cold storage. Full bins of dirty carrots go directly into the cold-storage room and wait in stacks for washing as time allows throughout the storage season (the other root vegetables—beets, daikon radishes, parsnips, rutabagas, and turnips—get the same treatment).

Rutabagas and parsnips are the final crops harvested at Food Farm. They use the Scott Viner for parsnips but not for rutabagas. Janaki said it's possible to modify the Scott Viner to harvest rutabagas, but he prefers to hand harvest into buckets rather than reconfiguring the machine. Cabbage root maggots are endemic to the farm, so rutabagas are extensively trimmed during harvest. Using the tops like handles, crew members cut off the root balls and all worm-damaged parts before cutting tops to about a quarter-inch (0.6 cm). The cut root surfaces might need to be lightly trimmed when packed, but they mostly do well in storage.

Although Food Farm has a space for indoor washing, they try to finish as much washing as possible outdoors in the fall. It's more time-efficient to wash veggies outdoors, and it's often too crowded in the indoor washing area early in the winter. Janaki tries to wash enough of each crop to cover sales until January, when there's more space for indoor washing.

Food Farm uses two primary washers depending on the crop. Round crops like potatoes, beets, and

The Food Farm harvest crew has a ball harvesting rutabagas into 5-gallon buckets! *Photo courtesy of Janaki Fisher-Merritt.*

rutabagas go through their brush washer. Carrots, parsnips, and daikon radishes go through their barrel washer. The farm has a hydraulic bin flipper to feed veggies into the wash line and a sorting table for grading cleaned crops. In the past, they felt it necessary to pressure wash veggies exiting the barrel washer to remove excess dirt and soil staining. Now, however, they send barrel-washed crops through the brush washer after passing the grading table. Janaki said the brush washer gives the carrots an attractive shine, especially by midwinter, when soil staining becomes an issue. He recently began sanitizing fall-washed carrots with SaniDate, a food-safe sanitizer, to reduce mold growth in storage. Washed root crops go into unlined bins with a single layer of burlap on the bottom to soak up excess moisture. Janaki prefers to let clean, wet veggies dry in the cold room overnight before placing a bin liner upside down over the outside of the bin.

Sales and Marketing

Winter sales at Food Farm make up about half of annual sales by weight. They organize a winter CSA offering monthly deliveries to customers starting in November and going through April, and CSA members can receive one or two boxes of loose veggies per month. During the winter, the farm packs and delivers about 275 boxes monthly. Food Farm also makes

The Food Farm wash line in action with parsnips on a sunny fall afternoon. *Photo courtesy of Janaki Fisher-Merritt.*

wholesale deliveries to Duluth twice a week until they run out of veggies not earmarked for the CSA. This varies by crop; they usually run out of wholesale carrots by February or March and cabbages by February. Janaki says he plans for losses in storage by growing and storing more than the farm needs. In an average year, he estimates that they lose 10 to 15 percent of crops like potatoes, carrots, and cabbages to spoilage. Food Farm has also found markets for lower-grade veggies. Some CSA customers purchase juice-grade carrots, which includes broken, large, and weirdly shaped roots. Food Farm also sells juice-grade veggies to a local kimchi/sauerkraut maker and a company making soup.

Diseases and Pests

Food Farm leaves all crop residues in the field and uses crop rotation to avoid problems with pests and diseases. They typically allow for at least four years between plantings from any given crop family. They also avoid putting crops that spoiled in storage back on the fields. Some spoiled crops, along with outright rejects, go to feed pigs at a nearby farm, while badly spoiled items go into a refuse pile that's located several hundred yards into the woods. Janaki said he doesn't want those spoiled items anywhere near the fields or the packing shed for fear of contamination.

In 2019, Food Farm experienced an infestation of white mold (*Sclerotinia sclerotiorum*) that caused heavy losses in stored carrots; Janaki estimated losing 40 to 50 percent of their 75,000-pound crop that year. The only available fungicide for white mold was an OMRI-listed soil additive containing another fungus (*Coniothyrium minitans*) that attacks the overwintering form of white mold. Janaki said the product is expensive and adds an additional step to field prep, but it appears to be helping reduce losses. To further reduce the effects of white mold on carrots, they've started washing a larger portion of the carrot crop in the fall. This allows them to sanitize the roots and hopefully lessen the impact of any white mold on their surfaces.

Advice for New Storage Farmers

Janaki advised farmers interested in winter storage, especially those already focused on summer and fall sales, to understand the ways it will change the flow of their farming season and year. On most farms, the transition from fall to winter is a tranquil time, a time to tie up the loose ends of the farming season before snow blankets the ground. For Janaki (and most other storage farmers, including me), fall is the busiest and most stressful time of the year. It's a challenge to harvest and process crops in time, and Janaki advised farmers to be ready for that rush. He also stressed that efficiency and ergonomics are especially important when harvesting and processing storage crops. Storage crops aren't as highly valued as many summer crops, and they also tend to be heavy and bulky. New storage farmers need to plan for efficiency in ways that won't destroy their bodies. Minimize the number of handling steps, use wheels whenever possible, and limit the weight workers must lift by hand.

Boldly Grown Farm

Owners: Amy Frye and Jacob Slosberg

Location: Bow, Washington; in the Skagit Valley about one hour north of Seattle

Farm size: 35 tillable acres with roughly 25 acres in storage crops

Field setup: Bed system spaced 60 inches (152.4 cm) center-to-center, either 350 feet (106.7 m) or 550 feet (167.6 m) long

Amount stored annually: 250,000 to 300,000 pounds (113,398 to 136,078 kg)

Signature storage crops: Radicchio, winter squash, carrots, beets, daikon and watermelon radishes, onions, kohlrabi, and cabbages; smaller amounts of shallots, garlic, and celeriac

Winter markets: Mostly wholesale (about 80 percent of annual business) but also a 325-member winter CSA and on-site farmstand

Summer markets: Wholesale and farmstand

Portion of annual sales from storage: Roughly 50 percent

Number of people working: Amy and Jacob full-time plus seven or eight full-time, year-round employees; an additional six or seven employees during peak harvest season

Boldly Grown Farm is a thriving operation growing a wide array of storage crops and offering a winter CSA. As I talked with the farmers and learned more about their history, I realized they are also a shining example of what hard work, intentional planning, and regional support for new farmers can create. Having started with 1 acre of crops at a local incubator farm in 2015, Amy Frye and Jacob Slosberg now cultivate 35 acres of their own land just eight years later. And from the beginning, they've focused on winter storage crops to fill gaps in the local food system and to stand out in the wholesale marketplace. Boldly Grown Farm shows that it's possible to both grow a farm business and scale quickly while focusing on storage crops. They're also growing a lot of radicchio! Although I wouldn't call radicchio a traditional storage crop, Amy and Jacob demonstrate that it can be brought into storage for a number of weeks and, in their mild maritime climate, be available most of the winter.

Journey to Wintertime Storage

After leaving the University of British Columbia's organic farm, where they met, Amy and Jacob moved to the Skagit Valley in Washington and started Boldly Grown Farm through an incubator program run by the nonprofit Viva Farms in 2015. From the start, winter storage helped them both stand out to distributors and spread their workload more evenly throughout the year. The Puget Sound Food Hub was a great help to Amy and Jacob as they grew their business. While expanding up to 12 acres at the incubator farm by 2021, Amy and Jacob were able to lease warehouse space operated by the Puget Sound Food Hub to serve as their wash-pack facility. This space was especially convenient because the Puget Sound Food Hub is one of their major distributors; they were able to wheel orders through the warehouse without driving anywhere. There, they also built a 500-square-foot cooler that they're still using today for off-farm storage.

During the 2021 season, Amy and Jacob were able to use their experience at the incubator farm to qualify for a Farm Service Agency loan and purchase their own land about 10 miles away. The property was an old dairy farm with 58 acres, a farmhouse, and about 50,000 square feet of barn space, most of which was unusable. In the latter half of 2021, they prepped 35 acres for production in 2022. Instead of demolishing the barns, they hired a salvaging company to deconstruct them and salvage materials. For example, they used the old concrete foundations as crushed rock for driveways and parking areas throughout the farm. They kept the nicest barn and renovated it to serve as the first on-farm wash-pack shed and to house the storage cooler.

Storage Spaces

At the time of this writing, Amy and Jacob store crops both on and off the farm. Boldly Grown Farm has continued to use its leased space at the Puget Sound Food Hub warehouse as backup storage, but our conversation mostly focused on the new storage facility on their property and their plans for the future.

They built the new storage cooler inside the barn they renovated to serve as the wash-pack shed. The building already had concrete floors and high ceilings, so they were able to make good use of the vertical space. Jacob bought a set of cheap used cooler panels that didn't meet the dimensions they wanted in a cooler, but they were able to incorporate them with some creative construction. They wanted their new cooler to

measure 32 by 34 feet with 15-foot ceilings, but the used wall panels were only 10 feet high and the ceiling panels too short at 24 feet. Jacob built a 5-foot pony wall with 2 × 6 lumber to support the 10-foot cooler walls. They filled the stud cavities with spray foam and sheathed both sides with plywood. For the ceiling, they sister-joined the too-short panels and supported them in the center with a horizontal beam running the length of the room. They added an overhead door to one wall and, voilà, had an on-site cooler space.

Amy and Jacob made a point of suggesting that farmers look for refrigeration specialists accustomed to projects in agricultural settings, because refrigeration units don't only need to keep the crops cold, but also need enough power to quickly remove field heat. They stressed that the ability to cool crops quickly is important to quality in long-term storage. Their target temperature in the cooler during the winter is 33°F (0.6°C). Jacob estimated their compressor power

Partway through the construction of an on-farm cooler. Amy and Jacob used framed pony walls to gain extra height and incorporate used cooler panels. *Photo courtesy of Amy Frye.*

in the range of 5 to 7 horsepower, and he said that the system pulls relatively little moisture from the air because of its large size. He also mentioned that having three-phase power has been a big help for running their cooling system because they're able to get more cooling power for the same amperage.

Amy and Jacob said that, ideally, humidity in the cooler would be over 95 percent, but in reality it hovers between about 70 and 90 percent. In their experience, the lower humidity is fine for everything except leafy crops like savoy cabbage and radicchio. They store moisture-sensitive root crops like carrots, beets, and radishes in macrobins with plastic bin liners set upside down over the tops. The liners are long enough to cover two bins at a time, and Amy and Jacob stressed the importance of allowing some airflow to the crops in storage.

Amy and Jacob used to store winter squash in the unheated warehouse space at the Food Hub, but they had issues maintaining proper temperatures there during cold snaps. Recently, they partnered with the National Resources Conservation Service and the local conservation district to add 2 inches of spray foam insulation to their on-farm pack shed. Now, they store both winter squash and onions there, where they're able to keep temperatures more stable around 45°F (7°C).

They also have a new arrangement with another nearby farm to use some of that farm's storage spaces seasonally. Amy said the other farm grows berries during the summer months, and its coolers would otherwise sit empty all winter. Instead, Boldly Grown Farm can utilize the spaces for storage crops, currently cabbages.

Materials Handling
Boldly Grown Farm stores the vast majority of its winter storage crops in plastic macrobins. Jacob estimated they currently have somewhere between 300

A view from the tractor: moving around macrobins full of winter squash in the twilight at Boldly Grown Farm. *Photo courtesy of Amy Frye.*

and 400 bins on the farm. To move these around, they've outfitted their tractor with forks; they have forklifts in the packing areas on and off the farm, and they use pallet jacks to maneuver in tight spaces. Whenever they hand-harvest, they use bulb crates to prevent themselves and their workers from lifting too much at a time.

They mentioned a new piece of handling equipment they're enjoying: a hydraulic bin trailer. This single-axle trailer mounts to their tractor's trailer tongue and, with hydraulics, can lower to nearly ground level to offload up to four bins at once. This action also allows crew members to load veggies into bins more easily, which saves time with crops like winter squash and cabbages. That said, the trailer doesn't have the same clearance as the normal harvest trailer and can't be driven over unharvested crops.

Amy and Jacob mentioned that when they farmed only a few acres at the incubator farm, they used smaller stacking crates made by a company called InterCrate for storage roots (see resources, page 237). The crates were both nestable and stackable, easy to clean, and small enough to lift by hand, making them ideal for small-scale storage and harvesting. They still use these smaller crates when hand-harvesting certain crops before dumping them into macrobins.

Harvesting and Curing

Even though Boldly Grown Farm does not attend markets or do a summer CSA, fall harvest is their busiest time of year. In addition to harvesting storage crops, October is the time when many of their fresh-marketed crops mature, and it's their biggest sales month to boot. To deal with this workload, the crew nearly doubles in size in October through December. Amy and Jacob said that their harvest schedule follows a clear progression as different crops mature and ripen, and they utilize the large crew to stay on track.

Onions are the first large storage planting they harvest, usually in late August to early September. Boldly Grown Farm hand-harvests their onions but first runs a tractor-mounted undercutter to make pulling easier. Amy and Jacob said they try to do some field curing, but this isn't a sure thing in their climate. Sometimes they're able to leave onions in the field for a few days after pulling, but other times they need to remove them right away before a big rain event soaks the drying bulbs. Crew members harvest onions into bulb crates until the crates are about half-full, and these crates are loaded onto a wagon and stacked in the greenhouse. After about a week the onions are fully dry and ready for processing. Crews use a roller topper to remove tops before loading dried onions into macrobins. This step is the main reason they harvest and cure onions in bulb crates: It would be too much hassle to move onions from macrobins into the topper by hand. Once in macrobins, the onions are stored in a hallway of the pack shed, which stays in the mid-40s °F (about 6 to 8°C) with low humidity for most of the winter. Amy and Jacob admit this isn't the ideal storage space for onions, but they value keeping onions dry over keeping them cold.

After onions, the crew harvests winter squash with the goal of getting them out of the field before the first frosts arrive, usually in late September or early October. They used to harvest squash into macrobins and leave those bins in the greenhouse for a couple weeks for stems to dry down and skins to harden. Now, Amy and Jacob field-cure the squash in windrows for one to two weeks. They clear all the squash from every third bed so they can drive the low-clearance hydraulic bin trailer down the empty beds, harvesting from both sides. There's no dedicated warm-dry storage area for the winter squash, but they plan to build one in the future. For now, they're using floorspace in the on-farm pack shed with temperatures set to the mid-40s °F.

After the first frosts come and go, the crew moves on to harvesting the root crops. If the weather is too warm, they may wait until temps cool down. During that time, they might harvest kohlrabi, which they prefer to get out earlier to prevent late-season cracking. Amy and Jacob said they want crops like carrots,

Jacob inspects stacks of drying onions in the greenhouse at Boldly Grown Farm. *Photo courtesy of Amy Frye.*

beets, winter radishes, and celeriac to experience some cold to develop flavor, but more importantly, they want to reduce the field heat they bring into the storage cooler. In their experience, the longer it takes to cool root crops after harvest, the more problems they see in storage. They try to harvest radishes first to minimize damage from root maggots. Celeriac are the last root crops out because they're so frost-tolerant. They used a Scott Viner for one season but upgraded to a new ASA-LIFT harvester for their root crops, including daikons and celeriac, which the Scott Viner could not harvest without damaging. They upgraded for several other reasons, too, including labor savings, greater reliability, and worker safety. The new harvester only requires two people to operate, so they can

keep up with other fall tasks during the root harvest. Macrobins full of dirty crops go into the on-farm cooler with bin liners set upside down over the tops. The crew eventually moves some of these crops to offsite storage when space on the farm gets tight.

I asked Amy and Jacob about their decisions to purchase and upgrade mechanical harvesters and what kind of improvements they've seen since. According to their records, it usually took one person eight hours to hand-harvest a three-row, 550-foot (168 m) bed of carrots after undercutting, and this usually produced about three macrobins. It took three people to run the Scott Viner—one running the harvester and two to pull a harvest trailer and direct carrots into bins. If everything was running

The new ASA-LIFT harvester fills a macrobin with freshly-harvested celeriac at Boldly Grown Farm. *Photo courtesy of Amy Frye.*

smoothly—which Jacob assured me wasn't always the case—they could harvest about four macrobins of carrots per hour. With the ASA-LIFT harvester they can do up to about six macrobins per hour with only two people. Amy and Jacob agree that the new harvester improves the quality of their veggies and also mitigates some risk, because they can afford to wait for good harvest weather.

Cabbages are the final crop harvested for long-term storage on Boldly Grown Farm. They are very tolerant to frost, and the quality of their heads isn't affected by cabbage root maggots. Jacob said the goal is to get them inside before they see any temperatures below 20°F (−7°C). When harvesting cabbages, crews trim off everything but the final wrapper leaves. Amy and Jacob stressed that they don't want loose cabbage leaves going into storage because they're bulky and

harbor more pathogens. They store both cabbages and kohlrabis in open, unlined macrobins in the 33°F (0.6°C) cooler. Low humidity in the cooler only causes issues with the leafier cabbages such as savoys, so they try to sell those first.

Boldly Grown Farm grows over 5 acres of radicchio! They plant them so they reach full size throughout the fall with a portion meant to last through the winter. Radicchio can withstand freezing temperatures down to about 25°F (−4°C) for a short period. Amy and Jacob opt to leave their radicchio in the field as long as possible, only bringing them into storage if it's absolutely necessary. In storage, the radicchio are more sensitive to desiccation than cabbages, and the crew often has to remove the outer layers while cleaning them for sales. Depending on their condition at harvest and the humidity in the cooler, radicchio last between four and eight weeks in storage. In addition to radicchio, Boldly Grown Farm overwinters kale as long as possible in the field, but they do not currently bring kale into storage.

Washing and Processing

Boldly Grown Farm stores most of their crops dirty. Their wash line includes a barrel washer, rinse conveyor, brush washer, and grading table all from the Pennsylvania-based AZS. They're currently loading veggies onto the wash line by hand via 3-gallon buckets, but Jacob assured me that a bin flipper and conveyor combo would be coming soon.

They use different washing tools depending on the crop. Winter squash and radishes go through the brush washer, whereas beets and carrots go through the barrel washer followed by the brush washer. They feel that the brush washer gives the veggies a clean, attractive shine that's harder to achieve with just the barrel washer or rinse conveyor alone. They mainly use the rinse conveyor during the summer for crops like peppers and bunched carrots.

After washing, crops exit the line onto a circular grading table where weird and broken pieces are removed. They're finding markets for some of these seconds, but the rest go into a compost pile. Amy and Jacob intend to mix the rotting vegetable matter with manure, let that decompose further, and eventually spread the finished mix back on the vegetable fields.

I found it interesting that the brush washer was the first piece of washing equipment they bought given that most storage farms use barrel washers. During their first three years of operation, they washed crops by presoaking them in harvest bins, hand spraying on a spray table, and polishing them in the brush washer. Amy and Jacob agreed that if a farm had just one piece of washing equipment, they would advise either a brush washer or rinse conveyor depending on the crop mix. Jacob said the brush washer is the simpler piece of equipment to set up and operate.

Marketing

Boldly Grown Farm has always been a wholesale-focused farm, but they've also been increasing their direct-market sales. They mainly sell through local distributors, including the Puget Sound Food Hub, which enables them to sell larger amounts of some crops, such as radicchio, than they could to a typical direct-market audience.

Amy joked that they are an accidental CSA farm. After starting the farm in 2015, friends and family in the Seattle area—where Jacob grew up—started asking about their veggies and how to get them. They decided to offer a small winter CSA to 12 interested people, and, in the 9 years since, they've grown to 325 members. The CSA doubled in size each year for a few years and has continued to grow with minimal advertising. CSA members receive prepacked boxes every other week in November through March and can choose from fifteen drop sites between Seattle (an hour south) and Bellingham (thirty minutes north), as well as on the farm. Now that the crazy transition to life at the new farm is winding down, Amy and Jacob say they'd like to make an intentional plan for how large or small they want the winter CSA to be. Currently, the CSA accounts for about 20 percent of their annual sales, but they're considering increasing that number.

In December 2022, Amy and Jacob opened a farm stand at their new farm location. They purchased a prefab building from a local shed company, purchased some used deli coolers, and set up a self-serve pay station to get the stand running. Their stand features their own produce, value-added products, beans, grains, and eggs, in addition to a wide assortment of goods from other local producers, including meat, dairy, breads, and more. They have a full-time farm stand and CSA coordinator to handle the day-to-day decision-making and logistics of those enterprises. They also grow a ¼-acre market garden to supply the farm stand with a wider array of veggies than they regularly grow for wholesale and the CSA. The farm stand has been a huge success in its first year of operation and is encouragingly popular with the local community.

Diseases and Pests

Boldly Grown Farm is currently managing against two main insect pests: cabbage root maggots (*Delia brassicae*) and wireworms (primarily *Limonius spp.* and *Selatosomus pruininus*). The cabbage root maggots are endemic to the area and bother any farmers growing cole crops. Amy and Jacob have to cover all their brassica root crops with row cover and rotate them from year to year. Without these precautions, their radishes, turnips, and rutabagas would be Swiss cheese. Jacob noted that it's unfortunate that soil conditions on organic and sustainable farms are exactly the conditions that root maggots thrive in: lots of cover crops and high organic matter. The wireworms are present at the new farm because, prior to 2021, the fields were pasture for fifteen to twenty years. They expect that with time and rotations, the wireworms won't be as problematic anymore.

Moving to their new farm relieved Amy and Jacob of some plant disease issues. At the incubator farm, they experienced a range of fungal pathogens affecting storage crops. The fields there had been in vegetable production for decades and, as such, had lots of time to build up hearty populations of fungi harmful to vegetable crops. Amy and Jacob recalled one particularly bad year when stored carrots turned to goo because of *Sclerotinia*. That happened to be the same year they tried wrapping stretch plastic over the plastic bin covers to create airtight seals. "Turns out," Jacob said, "vacuum-packing your carrots is a bad idea." They think they've avoided disease issues at the new farm location because vegetables have never been grown there before. They've also been trying an organic-certified product containing the beneficial fungus *Coniothyrium minitans* to keep *Sclerotinia* under control (see "Food Farm," page 219).

Amy and Jacob also shared advice they received from their HVAC installer, who often works with large potato farms. The installer said that reducing the airborne pathogen load in storage is important to reducing the incidence of storage rots. On that advice, they've talked about adding louvered air-intake fans to bring fresh air into their coolers. This, as Jacob added, could also become a method of cooling the spaces with outside air when weather conditions are right.

Advice for New Storage Farmers

Jacob advised that new storage farmers start small. He also talked about the importance of scaling growth with either more help from employees or with labor-saving equipment. As he said, you can't just add a couple of acres of carrots without adjusting the farm's capacity for weeding, harvesting, and processing them. Amy added that farmers need to think through the whole chain of events when getting into winter storage. She said that a large component of storage farming is materials handling—just moving heavy vegetables from place to place. The couple agreed that taking on well-reasoned debt is a likely part of scaling a storage farm, both for equipment and infrastructure.

Amy also advised farmers interested in winter storage to go into it with their eyes wide open. Storage farmers don't get much offseason—or any, depending on what else they're growing. She said farmers need to be ready for and accept that change in lifestyle. Along those lines, Amy and Jacob are trying to be intentional about giving themselves a bit of downtime. For example, they contract out the growing of their early transplants to give themselves a break in the spring. They admitted they haven't always succeeded in giving themselves enough time off, but they're trying to get better at it, especially as their kids, currently ages three and seven, get older. Amy stressed that taking actual time off isn't possible without long-term, trusted employees who can handle day-to-day operations around the farm.

Acknowledgments

This book could not have happened without the contributions of time, knowledge, and support from so many. First, thank you to those who mentored and encouraged me through the writing process since, as I'm happy to point out, writing is not my first language. This includes Andrew Mefferd, author and editor/publisher of *Growing for Market* magazine, who encouraged me to continue writing and to pursue this book project. This also includes the team at Chelsea Green Publishing, but particularly Fern Bradley, who mentored me through the process of submitting a book proposal, and Natalie Wallace, whose edits markedly improved my writing throughout the book.

Thank you to those who helped me in my journey toward farming, and storage farming in particular. This includes Brian Gronske, formerly of Groché Organic Farms in Little Suamico, Wisconsin, who served as my farmer mentor and first suggested I pursue winter storage. I must also thank the Mattila family, especially John and Diane, who offered incredible support to a budding fellow farmer, letting me farm on their land essentially for free and make alterations to their house for storage. So many friends helped me get Offbeet Farm off the ground; you know who you are, and I thank you.

I couldn't have written this book without help from many people who contributed their time, experience, and knowledge. I am particularly grateful to those farmers who took the time to be interviewed about their farms, look over rough drafts, and contribute photos from their operations: Sarah Hayward at North Farm in Chatham, Michigan; Jonathan Stevens at Jonathan's Farm in Winnipeg, Manitoba;

Jillian and Ross Mickens at Open Door Farm in Cedar Grove, North Carolina; Angus Baldwin and Holly Simpson Baldwin at West Farm in Jeffersonville, Vermont; Steve Pincus and Beth Kazmar at Tipi Produce in Evansville, Wisconsin; Janaki Fisher-Merritt at Food Farm in Wrenshall, Minnesota; Göran Johanson and Jan Goranson at Goranson Farm in Dresden, Maine; Amy Frye and Jacob Slosberg at Boldly Grown Farm in Bow, Washington; Bo Varsano and Marja Smets at Farragut Farm in Petersburg, Alaska; and Jesse Perkins at Mythic Farm in Blue Mounds, Wisconsin. Thanks to the other farmers who offered their experience and photos to the project: Allison Stawara at Partridge Creek Farm in Ishpeming, Michigan; Maria Papp at Bender Mountain Farm in Fairbanks, Alaska; Pete and Lynn Mayo at Spinach Creek Farm in Fairbanks, Alaska; Maggie Hallam at Cripple Creek Organics in Fairbanks, Alaska; Zoe Fuller at Singing Nettle Farm in Palmer, Alaska; Clara Coleman at Four Season Farm in Harborside, Maine; and Kevin Martin of Martin's Produce Supplies in Shippensburg, Pennsylvania. Thank you as well to the researchers and professionals who kindly offered their time and knowledge to a complete stranger: Irwin Goldman at the University of Wisconsin, Cindy Tong at the University of Minnesota, Chris Callahan at the University of Vermont Extension, and both Freddy Remolina and Julia DeGennaro at Store It Cold.

A lot of research went into this project, but there were a few works on which I relied heavily to build my framework of knowledge around winter vegetable storage. "The Commercial Storage of Fruits,

235

Vegetables, and Florist and Nursery Stocks," compiled by the USDA Agricultural Research Service with contributions from many authors, is a comprehensive handbook with a solid base of information for successfully storing a wide range of crops. I also credit Scott Sanford and John Hendrickson at the University of Wisconsin Extension for their guide "On-Farm Cold Storage of Fall-Harvested Fruit and Vegetable Crops," which informed my thinking and outlines regarding cold storage countless times.

Lastly, I give heartfelt thanks to my family, who supported me through starting farms and writing this book. Special thanks to my father, Phil Knapp, who became a talented photographer in his retirement. In addition to helping me build parts of the farm and guiding me with his wealth of knowledge, he came to Alaska for several weeks to take many of the photos featured in this book. I hope that someday I might possess a fraction of his giving spirit. This book would not exist without the steadfast support I received from my wife Danielle. She shouldered more burdens than was fair while I was consumed by building, farming, and writing by turns. Danielle offered invaluable guidance on the form and structure of the book, and she peeled my prone, despairing form off the floor more times than I care to count. I couldn't have done it without you. Thank you.

Resources

Josh Volk / Slow Hand Farm
www.slowhandfarm.com
Designs for hand carts.

Martin's Produce Supplies
www.martinsproducesupplies.com
Carries aluminum harvest conveyors.

Farm Hack
www.farmhack.org
Open-source designs for DIY farm tools, including a nifty barrel washer.

Nolt's Produce Supplies
www.noltsproducesupplies.net
Carries AZS washing equipment.

Market Farm Implement
www.marketfarm.com
Carries AZS washing equipment.

InterCrate
www.intercrate.com
Small stackable harvest crates.

CoolSeal USA
www.coolsealusa.com
Corrugated plastic boxes for CSA.

UVM Extension Agricultural Engineering Wall Dew Point Calculator
https://ageng.w3.uvm.edu/walls/index.html
A construction simulator that calculates the locations of dew points (if present) in specific wall designs and climate conditions.

Store It Cold
www.storeitcold.com
Website for the CoolBot.
Walk-in Cooler construction guide: www.storeitcold.com/construction-guide
CoolBot Sizing Chart: www.storeitcold.com/build-it/ac-selection
Sizing chart for properly sizing ACs with CoolBot systems. Note that many storage farms have cooling loads that exceed this chart.

National Cooperative Extension
https://farm-energy.extension.org/energy-answers-for-the-beginning-farmer-and-rancher
Video featuring Scott Sanford of the University of Wisconsin explaining cooling load calculations for on-farm coolers. This link goes to a page with many Cooperative Extension videos; find "How do I size a cooler for on-farm produce storage?" in the list. Alternatively, go to the YouTube page for "Energy Answers for the Beginning Farmer & Rancher" for this video and others.

UVM Extension Relative Humidity Table
https://blog.uvm.edu/cwcallah/files/2023/03/RH-Lookup-Table-Cold-and-Humid.pdf
A psychometric table for accurately measuring relative humidity in a cold-storage space with a sling psychrometer.

Active Air Humidifiers
www.hydrofarm.com/active-air
Commercial-grade centrifugal humidifiers.

Notes

Chapter 1. Choosing Crops and Varieties

1. Rachel S. Meyer, Ashley E. DuVal, and Helen R. Jensen, "Patterns and Processes in Crop Domestication: An Historical Review and Quantitative Analysis of 203 Global Food Crops," *New Phytologist* 196, no. 1 (October 2012): 29–48, https://doi.org/10.1111/j.1469-8137.2012.04253.x.

2. Melissa Bredow and Virginia K. Walker, "Ice-Binding Proteins in Plants," *Frontiers in Plant Science* 8 (December 2017): 2153, https://doi.org/10.3389/fpls.2017.02153.

3. Tortsten Nilsson, "Postharvest Handling and Storage of Vegetables," in *Fruit and Vegetable Quality*, ed. Robert L. Shewfelt and Bernhard Bruckner (Boca Raton, LA: CRC Press, 2000), 112–38.

4. Paul D. Hurd, Jr., E. Gorton Linsley, and Thomas W. Whitaker, "Squash and Gourd Bees (Peponapis, Xenoglossa) and the Origin of the Cultivated Cucurbita," *Evolution* 25, no. 1 (March 1971): 218–34, https://doi.org/10.2307/2406514; Patricia O'Brien, "The Sweet Potato: Its Origin and Dispersal," *American Anthropologist* 74, no. 3 (June 1972): 342–65, https://doi.org/10.1525/aa.1972.74.3.02a00070.

5. John Stolarczyk and Jules Janick, "Carrot: History and Iconography," *Horticulturae* 51, no. 2 (2011): 13–18.

6. N. P. Aksenova et al., "Regulation of Potato Tuber Dormancy and Sprouting," *Russian Journal of Plant Physiology* 60 (May 2013): 301–12, https://doi.org/10.1134/S1021443713030023.

7. Steven T. Koike, Peter Gladders, and Albert O. Paulus, *Vegetable Diseases: A Color Handbook* (Burlington, MA: Academic Press, 2007).

8. Kavita Sharma et al., "Importance of Growth Hormones and Temperature for Physiological Regulation of Dormancy and Sprouting in Onions," *Food Reviews International* 32, no. 3 (March 2016): 233–55, https://doi.org/10.1080/87559129.2015.1058820.

9. Johnny's Selected Seeds, *2023 Catalog* (Albion, ME: 2023), https://www.johnnyseeds.com/shop-johnnys/online-catalog-2023-pdf.html.

10. Organic Seed Alliance, *The Grower's Guide to Conducting On-farm Variety Trials* (Port Townsend, WA: 2017), https://seedalliance.org/wp-content/uploads/2018/03/Growers-guide-on-farm-variety-trials_FINAL_Digital.pdf.

Chapter 2. Harvest

1. "NWS Weather Forecast Offices," Climate, National Weather Service, https://www.weather.gov/wrh/climate.

Chapter 3. Postharvest Handling

1. Lars Mytting, *Norwegian Wood: Chopping, Stacking, and Drying Wood the Scandinavian Way*, trans. Robert Ferguson (New York: Abrams Image, 2015).

2. Nilsson, "Postharvest Handling and Storage."

3. Edward C. Lulai, "The Canon of Potato Science: 43. Skin-Set and Wound-Healing/Suberization," *Potato Research* 50 (July 2007): 387–90, https://doi.org/10.1007/s11540-008-9067-4.

4. Mike Bubel and Nancy Bubel, *Root Cellaring: Natural Cold Storage of Fruits & Vegetables*, 2nd ed., (North Adams, MA: Storey Publishing, 1991), 38.

5. Ruth Hazzard, Zara Dowling, and Amanda Brown, "From Field to Storage: High Quality Carrots," (proceedings of presentation at New England Vegetable and Fruit Conference, Manchester, NH, December 17–19, 2013), https://ag.umass.edu/sites/ag.umass.edu/files/pdf-doc-ppt/rhazzard_manchester_2013_from_field_to_storage_proceedings.pdf.

Chapter 4. Into Storage

1. Parsons and Henry Day, "Freezing Injury of Root Crops."

2. J. R. Stow, "Effects of Humidity on Losses of Bulb Onions (*Allium cepa*) Stored at High Temperature," *Experimental Agriculture* 11, no. 2, (April 1975): 81–87, https://doi.org/10.1017/S0014479700006517.

3. Sharma et al., "Importance of Growth Hormones"; R. C. Wright, J. I. Lauritzen, and T. M. Whiteman, "Influence of Storage Temperature and Humidity on Keeping Qualities of Onions and Onion Sets," *Technical Bulletin (United States Department of Agriculture)* 475 (May 1935).

4. Marita Cantwell, "Garlic: Recommendations for Maintaining Postharvest Quality," University of California Davis Postharvest Research and Extension Center, 2000, accessed October 20, https://postharvest.ucdavis.edu/produce-facts-sheets/garlic; Marita Cantwell, "Garlic," in *The Commercial Storage of Fruits, Vegetables, and Florist and Nursery Stocks*, ed. Kenneth C. Gross, Chien Yi Wang, and Mikal Saltveit (Washington, DC: US Department of Agriculture, Agricultural Research Service, 2016), 333–35; Ma Estela Vázquez-Barrios et al., "Study and Prediction of Quality Changes in Garlic cv. Perla (*Allium sativum* L.) Stored at Different Temperatures," *Scientia Horticulturae* 108, no. 2 (April 2006): 127–32; Gayle M. Volk, Kate E. Rotindo, and Walter Lyons, "Low-Temperature Storage of Garlic for Spring Planting," *HortScience* 39, no. 3 (June 2004): 571–73.

5. Howard W. Hruschka, Wilson L. Smith, and James E. Baker, "Reducing Chilling Injury of Potatoes by Intermittent Warming," *American Potato Journal* 46 (February 1969): 38–53, https://doi.org/10.1007/BF02868692; Nora Olsen, "Potato Dormancy Overview," University of Idaho College of Agricultural and Life Sciences, 2009, accessed October 20, 2023, https://www.uidaho.edu/-/media/UIdaho-Responsive/Files/cals/programs/potatoes/Storage/Dormancy-overview-2009.pdf; Trevor Suslow and Ron Voss, "Potato, Early Crop: Recommendations for Maintaining Postharvest Quality," University of California Davis Postharvest Research and Extension Center, 1998, accessed November 9, 2023, https://postharvest.ucdavis.edu/produce-facts-sheets/potato-early-crop.

6. David H. Picha, "Chilling Injury, Respiration, and Sugar Changes in Sweet Potatoes Stored at Low Temperature," *Journal of the American Society for Horticultural Science* 112, no. 3 (May 1987): 497–502, https://doi.org/10.21273/JASHS.112.3.497.

7. Stanley J. Kays, "Sweetpotato," in *The Commercial Storage of Fruits, Vegetables, and Florist and Nursery Stocks*, ed. Kenneth C. Gross, Chien Yi Wang, and Mikal Saltveit (Washington, DC: US Department of Agriculture, Agricultural Research Service, 2016), 566–70; L. J. Kushman, "Effect of Injury and Relative Humidity during Curing on Weight and Volume Loss of Sweet Potatoes during Curing and Storage," *HortScience* 10, no. 3 (June 1975): 275–77, https://doi.org/10.21273/HORTSCI.10.3.275; David H. Picha, "Weight Loss in Sweet Potatoes During Curing and Storage: Contribution of Transpiration and Respiration," *Journal of the American Society for Horticultural Science* 111, no. 6 (November 1986): 889–92, https://doi.org/10.21273/JASHS.111.6.889.

8. Jeffrey K. Brecht, "Pumpkins and Winter Squash," in *The Commercial Storage of Fruits, Vegetables, and Florist and Nursery Stocks*, ed. Kenneth C. Gross, Chien Yi Wang, and Mikal Saltveit (Washington, DC: US Department of Agriculture, Agricultural Research Service, 2016), 514–17.

9. Marita Cantwell and Trevor Suslow, "Pumpkin and Winter Squash: Recommendations for Maintaining Postharvest Quality," University of California Davis Postharvest Research and Extension Center, 2002, accessed October 21, 2023, https://postharvest.ucdavis.edu/produce-facts-sheets/pumpkin-winter-squash.

10. F. J. Francis and C. L. Thomson, "Optimum Storage Conditions for Butternut Squash," *Proceedings of the American Society for Horticultural Science* 88 (1965): 451–56.

11. James F. Thompson et al., *Commercial Cooling of Fruits, Vegetables, and Flowers, Revised Edition*, Publication 21567 (Oakland, CA: University of California, Agriculture and Natural Resources, 2008).

12. "Cooling Methods: Room Cooling," Postharvest Management of Vegetables, accessed March 26, 2024, https://www.postharvest.net.au/postharvest-fundamentals/cooling-and-storage/cooling-methods; Thompson et al., *Commercial Cooling*.

Chapter 5. The Business of Storage Farming

1. Jean C. Buzby et al., "Updated Supermarket Shrink Estimates for Fresh Foods and Their Implications for ERS Loss-Adjusted Food Availability Data," in *Economic Information Bulletin No. 155* (Washington, DC: US Department of Agriculture, Economic Research Service, 2016), https://www.ers.usda.gov /webdocs/publications/44100/eib-155.pdf?v=0.

2. Terhi Suojala-Ahlfors, "Effect of Harvest Time On Storage Loss and Sprouting in Onion," *Agricultural and Food Science in Finland* 10, no. 4 (January 2001): 323–33, https://doi.org/10.23986/afsci.5704.

3. I. L. Goldman, "Biennial Crops," in *Encyclopedia of Plant and Crop Science*, ed. Robert M. Goodman (New York, NY: Routledge, 2004), 122.

4. Jennifer D. Wetzel, "Winter Squash: Production and Storage of a Late Winter Local Food" (Master's thesis, Oregon State University, 2018), https:// ir.library.oregonstate.edu/concern/graduate_thesis _or_dissertations/1j92gd35b.

5. Ben Hartman, *The Lean Farm: How to Minimize Waste, Increase Efficiency, and Maximize Value and Profits with Less Work* (White River Junction, VT: Chelsea Green Publishing, 2015).

Chapter 6. Not Your Grandmother's Root Cellar

1. Johann Koenigsberger, "The Temperature of the Earth's Interior," *Scientific American*, May 25, 1907, https://www.scientificamerican.com/article /the-temperature-of-the-earths-inter.

Chapter 7. Planning a Modern Root Cellar

1. Scott Gibson, "How to Insulate a Foundation: Choosing the right material and the right technique for a house with a walkout basement," *Green Building Advisor*, June 8, 2015, https://www.greenbuilding advisor.com/article/how-to-insulate-a-foundation.

2. Martin Holladay, "Cold-Weather Performance of Polyisocyanurate," *Green Building Advisor*, September 4, 2015, https://www.greenbuildingadvisor.com/article /cold-weather-performance-of-polyisocyanurate.

3. Martin Holladay, "Fastening Furring Strips to a Foam-Sheathed Wall," *Green Building Advisor*, November 26, 2010, https://www.greenbuilding advisor.com/article/fastening-furring-strips-to-a -foam-sheathed-wall.

4. Scott A. Sanford and John Hendrickson, *On-Farm Cold Storage of Fall-Harvested Fruit and Vegetable Crops* (Madison, WI: Cooperative Extension of the University of Wisconsin–Extension, 2015), https:// fruit.webhosting.cals.wisc.edu/wp-content/uploads /sites/36/2017/02/On-Farm-Cold-Storage.pdf.

5. Sanford and Hendrickson, *On-Farm Cold Storage*; Thompson et al., *Commercial Cooling*.

Chapter 8. Climate Control

1. "EPA's Refrigerant Management Program: Questions and Answers for Section 608 Certified Technicians," Stationary Refrigeration, United States Environmental Protection Agency, last modified January 26, 2024, https://www.epa.gov/section608/epas -refrigerant-management-program-questions-and -answers-section-608-certified.

2. Rick Schuler, *CoolBot vs. Commercial Chilling Systems in Walk-in Coolers* (Ames, IA: Practical Farmers of Iowa, 2015), https://practicalfarmers.b-cdn.net /wp-content/uploads/2018/10/14.OE_.H.cooler -analysis1.pdf.

3. CDH Energy Corporation, *Evaluation of the COOLbot Low-Cost Walk-In Cooler Concept*, (Albany, NY: New York State Energy Research and Development Authority, 2009), https://storeitcold .com/wp-content/uploads/2020/03/NYSERDA -CoolBot-Report-May-09.pdf.

4. Schuler, *CoolBot vs. Commercial Chilling Systems*.

5. Chris Callahan, "Managing Humidity and Condensation in Coolers," *UVM Extension Ag Engineering, The University of Vermont*, March 1, 2023, https:// blog.uvm.edu/cwcallah/2023/03/01/ managing-humidity-and-condensation-in-coolers/.

6. Robert E. Hardenburg, Alley E. Watada, and Chien Yi Wang, eds., *The Commercial Storage of Fruits, Vegetables, and Florist and Nursery Stocks, Agricultural Handbook 66*: (Washington, DC: US Department of Agriculture, Agricultural Research Service, 2016).

7. Hardenburg, Watada, and Wang, *Commercial Storage*.

Part 3. Storage-Crop Compendium

1. Irwin L. Goldman and John Navazio, "History and Breeding of Table Beet in the United States," *Plant Breeding Reviews* 22 (October 2002): 357–88, https://doi.org/10.1002/9780470650202.ch7.

2. Franciszek Adamicki, "Beet," in *The Commercial Storage of Fruits, Vegetables, and Florist and Nursery Stocks*, ed. Kenneth C. Gross, Chien Yi Wang, and Mikal Saltveit (Washington, DC: US Department of Agriculture, Agricultural Research Service, 2016), 234–36.; Steven T. Koike, Peter Gladders, and Albert O. Paulus, *Vegetable Diseases: A Color Handbook* (Burlington, MA: Academic Press, 2007).

3. Dan Charles, "From Culinary Dud to Stud: How Dutch Plant Breeders Built Our Brussels Sprouts Boom," *National Public Radio*, October 30, 2019, https://www.npr.org/sections/thesalt/2019/10/30/773457637/from-culinary-dud-to-stud-how-dutch-plant-breeders-built-our-brussels-sprouts-bo.

4. Thomas Moore Whiteman, "Freezing Points of Fruits, Vegetables and Florist Stocks," in *Marketing Research Report No. 196* (Washington, DC: US Department of Agriculture, Agricultural Marketing Service, 1957).

5. Chien Yi Wang, "Chilling and Freezing Injury," in *The Commercial Storage of Fruits, Vegetables, and Florist and Nursery Stocks*, ed. Kenneth C. Gross, Chien Yi Wang, and Mikal Saltveit (Washington, DC: US Department of Agriculture, Agricultural Research Service, 2016), 62–67.

6. Melitta Weiss Adamson, *Food in Medieval Times* (New York: Greenwood Publishing Group, 2004).

7. Dan Buettner, *The Blue Zones Kitchen: 100 Recipes to Live to 100* (Washington, DC: National Geographic Society, 2019).

8. Koike, Gladders, and Paulus, *Vegetable Diseases*.

9. John Stolarczyk and Jules Janick, "Carrot: History and Iconography," *Horticulturae* 51, no. 2 (2011): 13–18.

10. Silvia Bruznican et al., "Celery and Celeriac: A Critical View on Present and Future Breeding," *Frontiers in Plant Science* 10 (January 2020): 1699, https://doi.org/10.3389/fpls.2019.01699.

11. Binda Colebrook, *Winter Gardening in the Maritime Northwest: Cool Season Crops for the Year-Round Gardener*, 5th ed. (Gabriola Island, Canada: New Society Publishers, 2012); Wang, "Chilling and Freezing Injury."

12. Marita Cantwell and Trevor Suslow, "Celery: Recommendations for Maintaining Postharvest Quality," University of California Davis Postharvest Research and Extension Center, 1998, https://postharvest.ucdavis.edu/produce-facts-sheets/celery.

13. Michelle Toratani, "Dokonjo Daikon: The Radish with the Fighting Spirit," in *Vegetables (Proceedings of the Oxford Symposium on Food and Cookery)* (Sheffield, U.K.: Prospect Books, 2009).

14. Richard L. Hassell, "Radish," in *The Commercial Storage of Fruits, Vegetables, and Florist and Nursery Stocks*, ed. Kenneth C. Gross, Chien Yi Wang, and Mikal Saltveit (Washington, DC: US Department of Agriculture, Agricultural Research Service, 2016), 524–26.

15. Rina Kamenetsky et al., "Environmental Control of Garlic Growth and Florogenesis," *Journal of the American Society for Horticultural Science* 129, no. 2 (March 2004): 144–51, https://doi.org/10.21273/JASHS.129.2.0144.

16. Takeomi Etoh, "Germination of Seeds Obtained From a Clone of Garlic, *Allium sativum* L.," *Proceedings of the Japan Academy, Series B* 59, no. 4 (1983): 83–87, https://doi.org/10.2183/pjab.59.83.

17. Einat Shemesh-Mayer and Rina Kamenetsky-Goldstein, "Traditional and Novel Approaches in Garlic (*Allium sativum* L.) Breeding," in *Advances in Plant Breeding Strategies: Vegetable Crops: Volume 8: Bulbs, Roots and Tubers*, ed. Jameel M. Al-Khayri, S. Mohan Jain, and Dennis V. Johnson (Cham, Switzerland, Springer Cham: 2021), 3–49.

18. Christian James, Violaine Seignemartin, and Stephen J. James, "The Freezing and Supercooling of Garlic (*Allium sativum* L.)," *International Journal of Refrigeration* 32, no. 2 (March 2009): 253–60, https://doi.org/10.1016/j.ijrefrig.2008.05.012.

19. Priit Põldma et al., "Influence of Different Winter Covers on the Yield of Garlic (*Allium sativum*) in Estonia," *Acta Horticulturae* 1268 (2020): 149–54, https://doi.org/10.17660/ActaHortic.2020.1268.19.

20. Harris Pearson Smith, George E. Altstatt, and Mills Herbert Byrom, "Harvesting and Curing of Garlic to Prevent Decay," *Bulletin (Texas Agricultural Experiment Station)* 651 (1944); A. M. El-Shabrawy et al., "Cultural Practices in Relation to Garlic Storage Diseases," *Assiut Journal of Agricultural Sciences* 18, no. 1 (1987): 5–18.

21. A. A. Kader, J. M. Lyons, and L. L. Morris, "Postharvest Responses of Vegetables to Preharvest Field Temperature," *HortScience* 9, no. 6 (December 1974): 523–27, https://doi.org/10.21273/HORTSCI.9.6.523.

22. Koike, Gladders, and Paulus, *Vegetable Diseases*.

23. Marita Cantwell, "Garlic: Recommendations for Maintaining Postharvest Quality," University of California Davis Postharvest Research and Extension Center, 2000, accessed October 20, 2023, https://postharvest.ucdavis.edu/produce-facts-sheets/garlic.; Marita Cantwell, "Garlic," in *The Commercial Storage of Fruits, Vegetables, and Florist and Nursery Stocks*, ed. Kenneth C. Gross, Chien Yi Wang, and Mikal Saltveit (Washington, DC: US Department of Agriculture, Agricultural Research Service, 2016), 333–35.; Ma Estela Vázquez-Barrios et al., "Study and Prediction of Quality Changes in Garlic cv. Perla (*Allium sativum* L.) Stored at Different Temperatures," *Scientia Horticulturae* 108, no. 2 (April 2006): 127–32.; Gayle M. Volk, Kate E. Rotindo, and Walter Lyons, "Low-Temperature Storage of Garlic for Spring Planting," *HortScience* 39, no. 3 (June 2004): 571–73.

24. "Storing Garlic," *Keene Garlic*, accessed November 9, 2023, https://keeneorganics.com/storing-garlic/; Petra Page-Mann, "Secrets of Storing Garlic," *Cornell Small Farms Program*, July 20, 2020, accessed October 4, 2023, https://smallfarms.cornell.edu/2020/07/secrets-of-storing-garlic/.

25. Vázquez-Barrios et al., "Study and Prediction of Quality Changes in Garlic."

26. Craig Reda, "Pre-Breeding of Kale (*Brassica oleracea* var. *acephala*)–Organic Adaptation and Shelf Life" (Master's thesis, Clemson University, 2022), https://tigerprints.clemson.edu/all_theses/3881.

27. Kader, Lyons, and Morris, "Postharvest Response of Vegetables."

28. Karin Albornoz and Marita Cantwell, "Fresh-Cut Kale Quality and Shelf-Life in Relation to Leaf Maturity and Storage Temperature," *Acta Horticulturae* 1141 (October 2016): 109–16, https://doi.org/10.17660/ActaHortic.2016.1141.11.

29. James W. Rushing, "Garlic," in *The Commercial Storage of Fruits, Vegetables, and Florist and Nursery Stocks*, ed. Kenneth C. Gross, Chien Yi Wang, and Mikal Saltveit (Washington, DC: US Department of Agriculture, Agricultural Research Service, 2016), 353–55.

30. Whiteman, "Freezing Points."

31. Koike, Gladders, and Paulus, *Vegetable Diseases*; Peter M.A. Toivonen and Charles Forney, "Kohlrabi," in *The Commercial Storage of Fruits, Vegetables, and Florist and Nursery Stocks*, ed. Kenneth C. Gross,

Chien Yi Wang, and Mikal Saltveit (Washington, DC: US Department of Agriculture, Agricultural Research Service, 2016), 377–78.

32. Maxine Saltonstall and Virginia Carroll, *First You Take A Leek* (North Clarendon, VT: Tuttle Publishing, 1969).

33. Whiteman, "Freezing Points."

34. Eugeniusz Kołota and Katarzyna Adamczewska-Sowińska, "The Effects of Flat Covers on Overwintering and Nutritional Value of Leeks," *Journal of Fruit and Ornamental Plant Research* 66, no. 1 (January 2007): 11–16.

35. Howard W. Hruschka, "Storage and Shelf Life of Packaged Leeks," in *Marketing Research Report No. 1084* (Washington, DC: United States Department of Agriculture, Agricultural Marketing Service, 1978).

36. Hruschka, "Packaged Leeks."

37. M. J. Havey, "Onion and other cultivated alliums," in Evolution of Crop Plants, 2nd ed., ed. J. Smartt and Norman Simmonds (Hoboken, NJ: Wiley, 1995), 344–50.

38. Rebecca Rupp, *How Carrots Won the Trojan War: Curious (but True) Stories of Common Vegetables* (North Adams, MA: Storey Publishing, 2011).

39. Robert Claude Wright, "Some Effects of Freezing on Onions," *United States Department of Agriculture: Department Circular* 415 (May 1927): 1–8.

40. Franciszek Adamicki, "Onion," in *The Commercial Storage of Fruits, Vegetables, and Florist and Nursery Stocks*, ed. Kenneth C. Gross, Chien Yi Wang, and Mikal Saltveit (Washington, DC: US Department of Agriculture, Agricultural Research Service, 2016), 406–10.; Kristina Mattson, "Kvalitet och lagring [Quality and storage]," in *Ekologisk grönsaksodling på friland [Organic vegetable growing in open fields]* (Jönköping, Sweden: Jordbruksverket [Swedish Board of Agriculture], 2015), https://webbutiken.jordbruksverket.se/sv/artiklar/p1013.html; Doyle A. Smittle and Bryan W. Maw, "Effects of Maturity and Harvest Methods on Storage and Quality of Onions," *HortScience* 23, no. 1 (February 1988): 141–43, https://doi.org/10.21273/HORTSCI.23.1.141; Peter Wright, D. G. Grant, and Christopher Michael Triggs, "Effects of onion (*Allium cepa*) plant maturity at harvest and method of topping on bulb quality and incidence of rots in storage," *New Zealand Journal of Crop and Horticultural Science*

29, no. 2 (June 2001): 85–91, https://doi.org / 10.1080/01140671.2001.9514166.

41. Mattson, "Kvalitet och lagring."

42. Wright, Grant, and Triggs, "Effects of Onion (*Allium cepa*)."

43. Linus U. Opara, "ONION: Post-Harvest Operation," in *INPhO – Post-harvest Compendium* (Food and Agriculture Organization of the United Nations: 2003), https://www.fao.org/3/av011e/av011e.pdf.

44. Katherine Downes, Gemma A. Chope, and Leon A. Terry, "Effect of Curing at Different Temperatures on Biochemical Composition of Onion (*Allium cepa* L.) Skin from Three Freshly Cured and Cold Stored UK-Grown Onion Cultivars," *Postharvest Biology and Technology* 54, no. 2 (November 2009): 80–86, https://doi.org/10.1016/j.postharvbio.2009.05.005.

45. Mattson, "Kvalitet och lagring."

46. Kalyani Gorrepati et al. 2017, "Curing of onion: A review," *Indian Horticulture Journal* 7, no. 1 (January –March 2017), 8–14, https://www.researchgate.net /publication/343205908_Curing_of_Onion _A_Review.

47. J. R. Stow, "Effects of Humidity on Losses of Bulb Onions (*Allium cepa*) Stored at High Temperature," *Experimental Agriculture* 11, no. 2, (April 1975): 81–87, https://doi.org/10.1017/S0014479700006517.

48. Adamicki, "Onion."

49. Koike, Gladders, and Paulus, *Vegetable Diseases*.

50. Stolarczyk and Janick, "Carrot: History and Iconography."

51. Ryosuke Munakata et al., "Molecular evolution of parsnip (*Pastinaca sativa*) membrane-bound prenyl-transferases for linear and/or angular furanocoumarin biosynthesis," *New Phytologist* 211, no. 1 (July 2016): 332–44, https://doi.org/10.1111/nph.13899.

52. Whiteman, "Freezing points"; Chester Swan Parsons and Richard Henry Day, "Freezing Injury of Root Crops: Beets, Carrots, Parsnips, Radishes, and Turnips," in *Marketing Research Report No. 866* (Washington, DC: US Department of Agriculture, Agricultural Marketing Service, 1970).

53. Peter M. A. Toivonen, "Parsnip," in *The Commercial Storage of Fruits, Vegetables, and Florist and Nursery Stocks*, ed. Kenneth C. Gross, Chien Yi Wang, and Mikal Saltveit (Washington, DC: US Department of Agriculture, Agricultural Research Service, 2016), 457–59.

54. B. B. Chubey and D. Gordon Dorrell, "Enzymatic Browning of Stored Parsnip Roots," *Journal of the American Society for Horticultural Science* 97, no. 1 (January 1972): 107–9, https://doi.org/10.21273 /JASHS.97.1.107.

55. Peter M. A. Toivonen, "The Reduction of Browning in Parsnips," *Journal of Horticultural Science* 67, no. 4 (1992): 547–51, https://doi.org/10.1080/00221589 .1992.11516282.

56. Koike, Gladders, and Paulus, *Vegetable Diseases*.

57. Stef de Haan and Flor Rodriguez, "Potato Origin and Production," in *Advances in Potato Chemistry and Technology*, 2nd ed., ed. Jaspreet Singh and Lovedeep Kaur (Cambridge, MA: Academic Press, 2016), 1–32.

58. Haan and Rodriguez, "Potato Origin and Production."

59. Trevor Suslow and Ron Voss, "Potato, Early Crop: Recommendations for Maintaining Postharvest Quality," University of California Davis Postharvest Research and Extension Center, 1998, accessed November 9, 2023, https://postharvest.ucdavis.edu /produce-facts-sheets/potato-early-crop.

60. Lewis Ralph Jones, "Frost Necrosis of Potato Tubers," *Research Bulletin (Agricultural Experiment Station of the University of Wisconsin)* 46 (October 1919), https:// books.google.com/books/about/Frost_Necrosis _of_Potato_Tubers.html?id=rxFGAQAAMAAJ.

61. Suslow and Voss, "Potato, Early Crop."

62. Reena Grittle Pinhero and Ricky Y. Yada, "Postharvest Storage of Potatoes," in *Advances in Potato Chemistry and Technology*, 2nd ed., ed. Jaspreet Singh and Lovedeep Kaur (Cambridge, MA: Academic Press, 2016), 283–314.

63. Steven B. Johnson, "Vine Killing," in *Commercial Potato Production in North America*, 2nd ed., ed. William H. Bohl and Steven B. Johnson (Livonia, MI: Extension Section of the Potato Association of America, 2010), 75–76, https://potatoassociation.org /wp-content/uploads/2014/04/A_Production Handbook_Final_000.pdf; Corné Kempenaar and Paul C. Struik, "The Canon of Potato Science: 33. Haulm Killing," *Potato Research* 50 (July 2007): 341–45, https://doi.org/10.1007/s11540-008-9082-5.

64. Edward C. Lulai, "The Canon of Potato Science: 43. Skin-set and Wound-healing/Suberization," *Potato Research* 50 (July 2007): 387–90, https://doi.org /10.1007/s11540-008-9067-4.

65. Ronald E. Voss, "Potato," in *The Commercial Storage of Fruits, Vegetables, and Florist and Nursery Stocks*, ed. Kenneth C. Gross, Chien Yi Wang, and Mikal Saltveit (Washington, DC: US Department of Agriculture, Agricultural Research Service, 2016), 506–10.

66. Howard W. Hruschka, Wilson L. Smith, and James E. Bake, "Reducing chilling injury of potatoes by intermittent warming," *American Potato Journal* 46 (February 1969): 38–53, https://doi.org/10.1007/BF02868692.; Nora Olsen, "Potato Dormancy Overview," *University of Idaho College of Agricultural and Life Sciences*, 2009, accessed October 20, 2023, https://www.uidaho.edu/-/media/UIdaho-Responsive/Files/cals/programs/potatoes/Storage/Dormancy-overview-2009.pdf.; Suslow and Voss, "Potato, Early Crop."; Voss, "Potato."

67. Haan and Rodriguez, "Potato Origin and Production."

68. Michael S. Reid and Harlan K. Pratt, "Effects of Ethylene on Potato Tuber Respiration," *Plant Physiology* 49, no. 2 (February 1972): 252–55, https://doi.org/10.1104/pp.49.2.252.

69. Robert K. Prange et al., "Using Ethylene as a Sprout Control Agent in Stored 'Russet Burbank' Potatoes," *Journal of the American Society for Horticultural Science* 123, no. 3 (May 1998): 463–69, https://doi.org/10.21273/JASHS.123.3.463.; Irena Rylski, Lawrence Rappaport, and Harlan K. Pratt, "Dual Effects of Ethylene on Potato Dormancy and Sprout Growth," *Plant Physiology* 53, no. 4 (April 1974): 658–62, https://doi.org/10.1104/pp.53.4.658.

70. Nora Olsen, Jeff Miller, and Phillip Nolte, "Diagnosis and Management of Potato Storage Disease," *University of Idaho Extension* CIS 1131 (July 2006), https://www.uidaho.edu/-/media/uidaho-responsive/files/extension/publications/cis/cis1131.pdf?la=en&rev=203125c5a7ba4a36bc68e4c7968344ab.

71. Melody Kroll, "New Study Revises Origins of the Humble Rutabaga," Biological Sciences, *University of Missouri College of Art and Science*, July 1, 2019, https://biology.missouri.edu/news/new-study-revises-origins-humble-rutabaga.

72. Åsa Holmgren et al., "Västgötskans 'rotabagge' blev amerikanskans 'rutabaga.' [West Götaland's 'rotabagge' became 'rutabaga' in American English]," *Institutet för Språk och Folkminnen [Institute for Language and Folklore]*, December 8, 2021, https://www.matkult.se/kal/dialektens-rotabagge-blev-varldskand.html.

73. Elias Wessén, *Våra ord: deras uttal och ursprung [Our words: their pronunciation and origins]*, (Stockholm, Sweden: Norstedts Akademiska Förlag, 1979).

74. D. Spaner, "Agronomic and Horticultural Characters of Rutabaga in Eastern Canada," *Canadian Journal of Plant Science* 82, no. 1 (January 2002): 221–24, https://doi.org/10.4141/P01-086.

75. Whiteman, "Freezing points."

76. Maria Cecilia do Nascimento Nunes, "Rutabaga," in *The Commercial Storage of Fruits, Vegetables, and Florist and Nursery Stocks*, ed. Kenneth C. Gross, Chien Yi Wang, and Mikal Saltveit (Washington, DC: US Department of Agriculture, Agricultural Research Service, 2016), 535–37.

77. Koike, Gladders, and Paulus, *Vegetable Diseases*.

78. Dan G. Bock et al., "Genome Skimming Reveals the Origin of the Jerusalem Artichoke Tuber Crop Species: Neither from Jerusalem nor an Artichoke," *New Phytologist* 201, no. 3 (February 2014): 1021–30, https://doi.org/10.1111/nph.12560.

79. Nalma Kosaric et al., "The Jerusalem Artichoke as an Agricultural Crop," *Biomass* 5, no. 1 (1984): 1–36, https://doi.org/10.1016/0144-4565(84)90066-0.

80. William Whitson, "Jerusalem Artichoke: True Seed Production," *Cultivariable* (blog), September 22, 2017, accessed November 16, 2023, https://www.cultivariable.com/jerusalem-artichoke-true-seed-production/.

81. Stanley J. Kays, "Jerusalem Artichoke," in *The Commercial Storage of Fruits, Vegetables, and Florist and Nursery Stocks*, ed. Kenneth C. Gross, Chien Yi Wang, and Mikal Saltveit (Washington, DC: US Department of Agriculture, Agricultural Research Service, 2016), 365–68.

82. Howard Wilfred Johnson, "Storage Rots of the Jerusalem Artichoke," *Journal of Agricultural Research* 43 no. 4 (August 1931): 337–52, https://archive.org/details/sim_journal-of-agricultural-research_1931-08-15_43_4/page/336/mode/2up?view=theater.

83. Kays, "Jerusalem Artichoke."

84. Patricia O'Brien, "The Sweet Potato: Its Origin and Dispersal," *American Anthropologist* 74, no. 3 (June 1972): 342–65, https://doi.org/10.1525/aa.1972.74.3.02a00070.

85. Tim Denham, "Ancient and Historic Dispersals of Sweet Potato in Oceania," *Proceedings of the National Academy of Sciences* 110, no. 6 (January 2013): 1982–83, https://doi.org/10.1073/pnas.1221569110.

86. Jenna Harburg, "Cumal to Kūmara: The Voyage of the Sweet Potato Across the Pacific," *Hohonu* 12, (September 2013): 6–10.

87. Alyse Whitney, "Yam vs. Sweet Potato: What's the Difference?," *Bon Appétit*, September 20, 2022, accessed November 12, 2023, https://www.bonappetit.com/story/difference-between-sweet-potato-and-yam.; Lex Pryor, "The Deep and Twisted Roots of the American Yam," *The Ringer*, November 24, 2021, https://www.theringer.com/2021/11/24/22798644/yam-sweet-potato-american-history.

88. "Sweetpotato Varieties," *North Carolina Crop Improvement Association*, accessed November 11, 2023, https://www.nccrop.com/varieties.php/8/Sweetpotato.

89. David H. Picha, "Weight Loss in Sweet Potatoes During Curing and Storage: Contribution of Transpiration and Respiration," *Journal of the American Society for Horticultural Science* 111, no. 6 (November 1986): 889–92, https://doi.org/10.21273/JASHS.111.6.889.

90. Picha, "Weight Loss in Sweet Potatoes"; David H. Picha, "Carbohydrate Changes in Sweet Potatoes During Curing and Storage," *Journal of the American Society for Horticultural Science* 112, no. 1 (January 1987): 89–92, https://doi.org/10.21273/JASHS.112.1.89.

91. Luis O. Duque et al., "A Win–Win Situation: Performance and Adaptability of Petite Sweetpotato Production in a Temperate Region," *Horticulturae* 8, no. 2 (February 2022): 172, https://doi.org/10.3390/horticulturae8020172.

92. David H. Picha, "Chilling Injury, Respiration, and Sugar Changes in Sweet Potatoes Stored at Low Temperature," *Journal of the American Society for Horticultural Science* 112, no. 3 (May 1987): 497–502, https://doi.org/10.21273/JASHS.112.3.497.

93. Stanley J. Kays, "Sweetpotato," in *The Commercial Storage of Fruits, Vegetables, and Florist and Nursery Stocks*, ed. Kenneth C. Gross, Chien Yi Wang, and Mikal Saltveit (Washington, DC: US Department of Agriculture, Agricultural Research Service, 2016), 566–70.

94. W. M. Walter and W. E. Schadel, "Structure and Composition of Normal Skin (Periderm) and Wound Tissue from Cured Sweet Potatoes," *Journal of the American Society for Horticultural Science* 108, no. 6 (November 1983): 909–14, https://doi.org/10.21273/JASHS.108.6.909.

95. Picha, "Carbohydrate Changes in Sweet Potatoes."

96. Kays, "Sweetpotato."

97. Kays, "Sweetpotato"; L. J. Kushman, "Effect of Injury and Relative Humidity during Curing on Weight and Volume Loss of Sweet Potatoes during Curing and Storage," *HortScience* 10, no. 3 (June 1975): 275–77, https://doi.org/10.21273/HORTSCI.10.3.275.; Picha, "Weight loss in sweet potatoes."

98. C. F. Abrams, Jr., "Depth Limitations on Bulk Piling of Sweet Potatoes," *Transactions of the ASAE* 27, no. 6 (1984):1848–53, https://doi.org/10.13031/2013.33056.

99. Kays, "Sweetpotato."

100. H. Reiner, W. Holzner, and R. Ebermann, "The Development of Turnip-Type and Oilseed-Type Brassica Rapa Crops from the Wild-Type in Europe–An Overview of Botanical, Historical and Linguistic Facts," *Rapeseed Today and Tomorrow* 4 (1995):1066–69.

101. Rebecca Rupp, "The Vegetable that Terrorized Romans and Helped Industrialize England," *National Geographic*, May 8, 2014, https://www.nationalgeographic.com/culture/article/the-vegetable-that-terrorized-romans-and-helped-industrialize-england.

102. Whiteman, "Freezing Points."

103. Penelope Perkins-Veazie, 2016, "Turnip," in *The Commercial Storage of Fruits, Vegetables, and Florist and Nursery Stocks*, ed. Kenneth C. Gross, Chien Yi Wang, and Mikal Saltveit (Washington, DC: US Department of Agriculture, Agricultural Research Service, 2016), 590–91.

104. Koike, Gladders, and Paulus, *Vegetable Diseases*.

105. Ryan Pankau, "The History of Squash," *The Garden Scoop* (blog), *University of Illinois Urbana-Champaign, College of Agriculture, Consumer & Evironmental Sciences, Illinois Extension*, November 22, 2017, accessed November 19, 2023, https://extension.illinois.edu/blogs/garden-scoop/2017-11-22-history-squash.

106. Ann D. Roberts, *Alaska Gardening Guide: Alaska Vegetables for Northern Climates* (Anchorage, AK: Publishing Consultants, 2000).

107. "Heaviest pumpkin," *Guiness World Records*, 2023, accessed November 21, 2023. https://www.guinness worldrecords.com/world-records/heaviest-pumpkin.

108. Lacy Porter McColloch, "Alternaria Rot Following Chilling Injury of Acorn Squashes," in *Marketing Research Report No. 518* (Washington, DC: US Department of Agriculture, Agricultural Marketing Service, 1962).

109. McColloch, "Alternaria Rot Following Chilling"; Franklin D. Schales and F. M. Isenberg, "The Effect of Curing and Storage on Chemical Composition and Taste Acceptability of Winter Squash," *Proceedings of the American Society for Horticultural Science* 83 (1963): 667–74.

110. B. T. Hawthorne, "Age of Fruit at Harvest Influences Incidence of Fungal Storage Rots on Fruit of *Cucurbita maxima* D. Hybrid 'Delica,'" *New Zealand Journal of Crop and Horticultural Science* 18, no. 2–3 (1990): 141–45, https://doi.org/10.1080/01140671 .1990.10428085.

111. P. W. Sutherland and I. C. Hallett, "Anatomy of Fruit of Buttercup Squash (*Cucurbita maxima* D.) Surface, Cuticle, and Epidermis," *New Zealand Journal of Crop and Horticultural Science* 21, no. 1 (1993): 67–72, https://doi.org/10.1080/01140671 .1993.9513748.

112. Hawthorne, "Age of Fruit at Harvest."

113. Megan Robertson, "Tips for Avoiding Storage Rot in Winter Squash," *Growing for Market Magazine*, November 1, 2020, https://growingformarket.com /articles/tips-avoiding-storage-rot-winter-squash.

114. Rosa Marina Arvayo-Ortiz, Sergio Garza-Ortega, and Elhadi M. Yahia, "Postharvest Response of Winter Squash to Hot-water Treatment, Temperature, and Length of Storage," *HortTechnology* 4, no. 3 (July 1994): 253–55, https://doi.org/10.21273 /HORTTECH.4.3.253.; B. T. Hawthorne, "Effects

of cultural practices on the incidence of storage rots in *Cucurbita* spp," *New Zealand Journal of Crop and Horticultural Science* 17, no. 1 (1989): 49–54, https://doi.org/10.1080/01140671.1989.10428009.

115. Jeffrey K. Brecht, "Pumpkins and Winter Squash," in *The Commercial Storage of Fruits, Vegetables, and Florist and Nursery Stocks*, ed. Kenneth C. Gross, Chien Yi Wang, and Mikal Saltveit (Washington, DC: US Department of Agriculture, Agricultural Research Service, 2016), 514–17.

116. Schales and Isenberg, "The Effect of Curing."

117. Ethan Grundberg, "2019 Kaboch Squash Variety Trial," Eastern New York Commercial Horticulture, Cornell Cooperative Extension, powerpoint, https://rvpadmin.cce.cornell.edu/uploads/doc_865 .pdf.; Jennifer D. Wetzel, "Winter Squash: Production and Storage of a Late Winter Local Food" (Master's thesis, Oregon State University, 2018), https://ir.library.oregonstate.edu/concern /graduate_thesis_or_dissertations/1j92gd35b.

118. Virgil Severns and Anne Severns, *Winter Squash and Pumpkins for Northern Gardens* (Fairbanks, AK: A. F. Farmer, 2018).

119. Vuvu D. Manseka and James R. Hicks, "Postharvest Factors Affecting Quality of Butternut Squash during Storage," *HortScience* 31, no. 4 (August 1996): 599g–600, https://doi.org/10.21273/HORTSCI .31.4.599g.

120. Brecht, "Pumpkins and Winter Squash."

121. F. J. Francis and C. L. Thomson, "Optimum Storage Conditions for Butternut Squash," *Proceedings of the American Society for Horticultural Science* 88 (1965): 451–56.

122. Brecht, "Pumpkins and Winter Squash."

123. Brecht, "Pumpkins and Winter Squash."

Part 4. Storage Farm Profiles

1. Steven T. Koike, Peter Gladders, and Albert O. Paulus, *Vegetable Diseases: A Color Handbook* (Burlington, MA: Academic Press, 2007).

Index

Note: Page numbers in *italics* indicate photographs and illustrations; page numbers followed by *t* indicate tables.

About the Author

Martha Mintz

S am grew up near the forested shores of Lake Superior with little exposure to farm life. After earning degrees in physics and chemistry and beginning work as an engineer, Sam caught the farming bug while working on a research farm in Sweden. He worked on vegetable farms in Sweden, Alaska, Michigan, and Wisconsin before starting his own Root Cellar Farm in the Upper Peninsula of Michigan to earn extra income while completing a master's degree in plant ecology. In 2020, Sam moved to Fairbanks, Alaska, where he built Offbeet Farm from scratch in a patch of boreal forest. Offbeet Farm provides vegetables to the Fairbanks community during the long, dark Alaskan winters through a winter-only community supported agriculture (CSA) program and sales to local groceries. Owing to his scientific background, he is a self-professed nerd when it comes to energy-efficiency, farm tools, and spreadsheets. In his free time, Sam is an avid cross-country skier and musician.

the politics and practice of sustainable living

CHELSEA GREEN PUBLISHING

Chelsea Green Publishing sees books as tools for effecting cultural change and seeks to empower citizens to participate in reclaiming our global commons and become its impassioned stewards. If you enjoyed reading *Beyond the Root Cellar*, please consider these other great books related to farming and food systems.

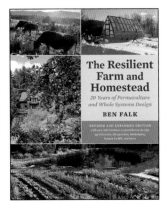

THE RESILIENT FARM AND HOMESTEAD, REVISED AND EXPANDED EDITION
20 Years of Permaculture and Whole Systems Design
BEN FALK
9781645021100
Paperback

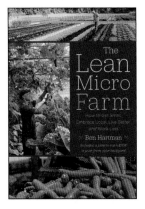

THE LEAN MICRO FARM
How to Get Small, Embrace Local, Live Better, and Work Less
BEN HARTMAN
9781645022046
Paperback

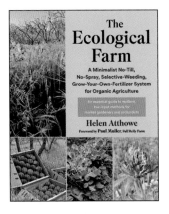

THE ECOLOGICAL FARM
A Minimalist No-Till, No-Spray, Selective-Weeding, Grow-Your-Own-Fertilizer System for Organic Agriculture
HELEN ATTHOWE
9781645021810
Paperback

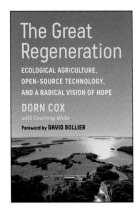

THE GREAT REGENERATION
Ecological Agriculture, Open-Source Technology, and a Radical Vision of Hope
DORN COX WITH COURTNEY WHITE
9781645020677
Paperback

CHELSEA GREEN PUBLISHING
the politics and practice of sustainable living

For more information,
visit **www.chelseagreen.com**.

the politics and practice of sustainable living

CHELSEA GREEN PUBLISHING

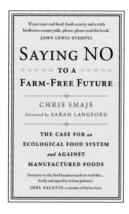

SAYING NO TO A FARM-FREE FUTURE
*The Case for an Ecological Food System
and Against Manufactured Foods*
CHRIS SMAJE
9781915294166
Paperback

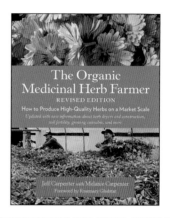

THE ORGANIC MEDICINAL HERB FARMER, REVISED EDITION
How to Produce High-Quality Herbs on a Market Scale
JEFF CARPENTER WITH MELANIE CARPENTER
9781645021124
Paperback

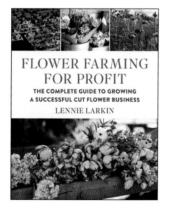

FLOWER FARMING FOR PROFIT
*The Complete Guide to Growing a
Successful Cut Flower Business*
LENNIE LARKIN
9781645021766
Paperback

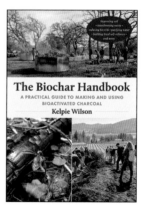

THE BIOCHAR HANDBOOK
*A Practical Guide to Making
and Using Bioactivated Charcoal*
KELPIE WILSON
9781645022305
Paperback

CHELSEA GREEN PUBLISHING

the politics and practice of sustainable living

For more information,
visit **www.chelseagreen.com.**